Estimating Costs of Air Pollution Control

William M. Vatavuk

T0225643

CRC Press
Taylor & Francis Group
Boca Raton London New York

CRC Press is an imprint of the
Taylor & Francis Group, an **informa** business

First published 1990 by Lewis Publishers

Published 2019 by CRC Press
Taylor & Francis Group
6000 Broken Sound Parkway NW, Suite 300
Boca Raton, FL 33487-2742

First issued in paperback 2020

ISBN 13: 978-0-367-58016-2 (pbk)
ISBN 13: 978-0-87371-142-5 (hbk)

Visit the Taylor & Francis Web site at
http://www.taylorandfrancis.com

and the CRC Press Web site at
http://www.crcpress.com

Library of Congress Cataloging-in-Publication Data

Vatavuk, William M.
  Estimating costs of air pollution control/by William M. Vatavuk.
    p. cm.
  Includes bibliographical references.
    1. Air—purification—equipment and supplies—Costs. 2. Air—
Purification—Estimates. I. Title.
TD889.V37   1990
628.5′3′028—dc20   89-13789
ISBN 0-87371-142-4

Without her patience, encouragement—and gentle prodding—this volume would never have made it to the printers. Therefore, I gratefully dedicate this book to Betsy, my wife.

Preface

Ever find yourself in this situation? You've been told to estimate the cost of controlling air pollutant emissions from sources at your plant. This estimate has to be accurate *and* it has to be made quickly. What do you do? Start phoning equipment vendors? Hire a consultant? Perhaps. But before you do, you should *read this book*.

In these pages is all the information that you—manager, engineer, or other technical professional—would need to select, size, and estimate "budget/study"-level capital and annual costs for a variety of air pollution control equipment. This equipment includes wet scrubbers, carbon adsorbers, and other "add-on" devices. This book also deals with such nonstack controls as wet dust suppression systems and flue gas desulfurization systems. The costs are current (1988 or 1989 dollars) and are mainly presented in equational form for ease of computerization and updating. Finally, several detailed example problems are included to illustrate the sizing and costing procedures.

This book is not just for technical personnel, however. The material is easy to grasp and use. I believe that anyone with an air pollution control background can follow and apply the procedures and data herein. Through all, I have tried not to lose sight of my goal: to help others in the air pollution control field develop sound, defensible (within ±30%) cost estimates with a minimum of time and effort.

William M. Vatavuk
Durham, North Carolina

Acknowledgments

I wish to thank the firms who contributed to this book by providing valuable data and advice. In alphabetical order, they are:

- Adams Industrial Sales (High Point, NC)
- Air Plastics, Inc. (Cincinnati, OH)
- Calvert Environmental Equipment (San Diego, CA)
- Chet Adams Co. (Cary, NC)
- Croll-Reynolds, Inc. (Westfield, NJ)
- Douglas Martin & Associates (Emmaus, PA)
- Edwards Engineering Corp. (Pompton Plains, NJ)
- Environmental Products Sales, Inc. (Raleigh, NC)
- Hoffman & Hoffman, Inc. (Raleigh, NC)
- Hoyt Corp. (Westport, MA)
- Intertel Corp. (Englewood, CO)
- Johnson March Corp. (Ivyland, PA)
- Koch Engineering (Wichita, KS)
- L. R. Gorrell Co. (Raleigh, NC)
- Multi-Duti Manufacturing (Baldwin Park, CA)
- Murphy-Rodgers, Inc. (Huntington Park, CA)
- R. S. Means Co. (Kingston, MA)
- Salem Industries, Inc. (South Lyon, MI)
- Screw Conveyor Corp. (Hammond, IN)
- Skilcraft Fiberglas Corp. (Glendale, AZ)
- Sur-Lite Corp. (Santa Fe Springs, CA)
- Tigg Corp. (Pittsburgh, PA)
- Universal Air Precipitator Corp. (Monroeville, PA)
- Vanaire, Ltd. (Louisville, KY)
- W. W. Sly Co. (Cleveland, OH)
- Zurn Constructors, Inc. (Tampa, FL)
- Zurn Industries, Inc., Air Systems Division (Birmingham, AL)

Several other firms (who requested anonymity) also supplied very useful information. I am very grateful to them as well. In addition, a very special group of people took the time and effort to review portions of the manuscript. Each of these reviewers is a recognized expert in his field. This book is much better, thanks to their valuable — and much valued — input. Their names and affiliations are:

- Dr. Howard Hesketh (Southern Illinois University)
- Mr. Hank McDermott (Chevron Corp.)
- Mr. Jay Matley (*Chemical Engineering* magazine)
- Mr. Gerald Schrubba (Salem Industries, Inc.)
- Mr. Kenneth Schifftner (Calvert Environmental Equipment)

Two other experts also reviewed parts of the manuscripts; however, they have asked to remain anonymous. I am deeply indebted to each of these gentlemen.

Finally, my hat is off to Ed Lewis, my publisher, and Robin Berry, my editor. Their sound advice and creative suggestions made this book easier to write, easier to read, and, most important, easier to *use*.

William M. Vatavuk is a senior chemical engineer with the Office of Air Quality Planning and Standards (OAQPS), U.S. Environmental Protection Agency, where he has been employed since 1970. For 15 of those 20 years he has specialized in air pollution control cost analysis, work that has supported the setting of national emission and air quality standards. A recognized national expert in this field, he has published over 40 technical articles on cost analysis–related topics for such magazines as *Chemical Engineering* and the *Journal of the Air Pollution Control Association*. He has also received two EPA bronze medals for his work in this area.

A graduate of Youngstown State University (Bachelor of Engineering, 1969), he is a registered professional engineer in North Carolina, a charter member of the OAQPS Human Resources Council, and twice past president of the North Carolina branch of the U.S. Public Health Service (USPHS) Commissioned Officers Association. A USPHS Commander, Mr. Vatavuk lives with his wife and son in Durham, North Carolina.

Contents

Estimating Costs of Air Pollution Control

CHAPTER 1

The Bottom Line

Money does not pay for anything, never has, never will.
Albert Jay Nock

Americans are impressed by *bigness*: big homes, cars, and bank balances; big accomplishments; big reputations. No wonder such government-generated figures as the balance of trade and budget deficits impress us. By their very magnitude, they are guaranteed to grab (and keep) our attention. Moreover, we are typically an impatient folk too busy to be bothered by details. We want to get to the "bottom line," be it the plumber's bill or the defense budget.

This chapter will deal with another sort of "bottom line"—the cost of controlling air pollution in the United States today. As we might expect, this is by no means a trivial expenditure. For instance, one projection for air pollution control total investment between 1981 and 1990 was $102 billion.[1] The word "estimate" is operative here, however, for exact control cost figures have not been compiled—if indeed they ever can be. Nonetheless, these estimates provide, at worst, general indicators of the cost impact of air pollution control. At best, they give us statistically sound data that approach the accuracy of a thorough census.

Three federal reports have provided such aggregated cost data: (1) *The Cost of Clean Air and Water* (U.S. Environmental Protection Agency), (2) "Plant and Equipment Expenditures by Business for Pollution Abatement" (U.S. Department of Commerce, Bureau of Economic Analysis), and (3) *Current Industrial Reports: Pollution Abatement Costs and Expenditures* (U.S. Department of Commerce, Bureau of the Census). Because these reports differ in objectives and readership, the results they show—and the methods they use to obtain them—differ as well. In the rest of this chapter, we'll take a closer look at each report, comparing and contrasting each with its rivals. Finally, we'll draw selective data from these reports and construct national air pollution control cost estimates of our own.

EPA's EFFORT: *THE COST OF CLEAN AIR AND WATER*

Those who drafted the bill that became the Clean Air Act (in December 1970) expected air pollution control to be a costly proposition. It only seemed reasonable, therefore, to give the public periodic reports on control's price tag. Section 312(a) of the act required these cost studies:

> [T]he Administrator shall make a detailed estimate of the cost of carrying out the provisions of this Act; a comprehensive study of the cost of program implementation by affected units of government; and a comprehensive study of the economic impact of air quality standards on the Nation's industries, communities, and other contributing sources of pollution, including an analysis of the national requirements for and the cost of controlling emissions to attain such standards of air quality as may be established pursuant to this Act or applicable State law. The Administrator shall submit. . .[these] studies annually. . . .[2]

From its inception, *The Cost of Clean Air and Water* ("COCAW") included only those costs associated with *federal* actions resulting from implementation of the Clean Air Act. Other control costs – those required by independent state or local regulations or voluntarily incurred by the pollution sources themselves – were not considered. For the most part, these costs were based on compliance with EPA-approved State Implementation Plans (SIP) (which incorporate rules for regulating emission sources within state or territory boundaries) and EPA-promulgated regulations for stationary sources (such as New Source Performance Standards [NSPS] and National Emission Standards for Hazardous Air Pollutants [NESHAP]) and for mobile sources. (*Note:* In later years, the report also contained analogous costs for complying with the 1972 Clean Water Act, but these results were determined and presented separately.)[3]

The 1984 COCAW is typical. In this report, all costs were expressed in 1981 dollars, the "base" year. Overall, these estimates covered two periods: 1970–81 (historical) and 1982–90 (projected). Two kinds of expenditures were shown: *Capital* costs – the investment associated with the purchase and installation of pollution control equipment – and *Annual* (or "Annualized") costs, which included costs for operating and maintaining the equipment ("O&M" costs) and such capital-related charges as depreciation and interest. In addition, where applicable, the annual costs incorporated *credits* for product(s) or energy recovered by a control system.[4]

Interestingly, the report used the same computer model–based procedure to calculate both the historical and projected costs – that is, for each industry or other "source category," cost functions were derived from data in EPA reports or other references. Usually nonlinear, these functions related cost (capital, annual, or recovery credit) to the size of a surrogate (or "model") facility in that source category. The sizes of these surrogate facilities were selected, in turn, so as to represent the *distribution* of facility sizes in the category.

For instance, consider a source category with 30 facilities ranging in capacity from 10 million to 40 million lb/year, with a total capacity of 440 million lb/year. One might select three surrogate facility sizes – 10, 20, and 40 million lb/year – and apportion the total category capacity among those sizes. The per-plant capital and annual control costs would be calculated for each surrogate facility size. Next, these surrogate facility costs would be multiplied by the number of facilities of that size. Finally, one would sum these products to obtain the total capital and annual costs for that source category.

This was a simplified example. Some source categories, such as steam-electric utilities, contain hundreds of facilities encompassing a wide range of capacities. Moreover, control costs depend on variables other than just capacity. Still, the authors of COCAW considered this "surrogate" approach to be accurate enough for purposes of making national control cost estimates.

The control cost functions were the foundation for this estimating procedure. Each type of control system applied to a source category had a unique set of functions. Question: "Who or what determined the selection of the control system?" Answer: "The type of regulation that the source had to meet – NSPS, SIP, etc." To meet a particulate SIP control level, one might select a low-energy wet scrubber, for example, while to comply with a more stringent NSPS, the choice might be a fabric filter. Because most source categories had facilities in several states, the computer model was programmed to select a representative state regulation.[5]

But the model didn't stop there. It also contained historical and estimated future *growth rates*. The model used these rates to compute the facility "population" for each category on a year-by-year basis. Of course, not every facility in existence would likely install controls in the first year the regulation was in effect. For this reason, "compliance schedules" were developed to reflect the *fraction* of the existing facilities that would comply with the regulation in any given year. Finally, the computer model included an algorithm for calculating the cost of replacing a control system once it wore out.[6]

However, the model was not used for all source categories. For some categories, the authors estimated and entered in "exogenous" costs – costs that they had obtained from EPA reports and other reliable references. In such cases, the source categories may have been too complex for use of the model. In other cases, detailed cost studies may have already been done (e.g., for the "Mobile Sources" category). In any event, these exogenous costs had the virtue of being consistent with other, publicly available studies, while we probably couldn't say the same for costs generated by the computer model.[7]

So much for the estimating procedure. How large were the costs reported in COCAW? That question is answered in Table 1.1. This table shows capital and annual costs for 14 source categories. (The report showed costs for 25, but combined some of the smaller categories into larger ones.) Costs for each category are shown for 1981 and the years 1981–90. (The report also presented costs for the periods 1970–78, 1979–81, and 1979–84, but these data are not as relevant to this discussion.)

These costs reflected the expenditures incurred to control existing and new sources in 1981 and 1981–90, *given what was known about those categories in 1981 and what was projected to occur in those categories during the next nine years*. The costs in the original report were expressed in 1981 dollars, but as

Table 1.1 Summary of 1984 *Cost of Clean Air and Water* Air Control Costs

Source Category	Capital (million $)		Annual (million $)	
	1981	1981-90	1981	1981-90
Stationary Sources				
Electric utilities	3,120	32,450	9,310	119,600
Other fuels and energy	150	1,050	240	3,320
Food processing	40	1,140	190	2,100
Chemicals	80	1,710	630	6,680
Construction materials	60	1,030	370	4,430
Metals	370	4,350	2,500	31,670
Softgoods	90	590	450	4,860
Manufacturing	150	1,280	150	2,840
Services	0	510	0	580
Municipal waste incineration	10	70	30	310
Other industrial	670	9,780	3,380	36,360
Subtotal:	4,740	53,960	17,250	212,740
Mobile Sources	5,730	73,720	7,560	100,690
Government Expenditures	0	0	760	6,580
TOTALS:[a]	10,500	128,000	25,600	320,000

Source: Cost of Clean Air and Water: Report to Congress. Washington, DC: U.S. Environmental Protection Agency, Office of Policy Analysis, May 1984. (All costs shown have been escalated to 1987 dollars.)
[a]Totals have been rounded to three places.

that measure has little meaning today, we have escalated them to 1987 dollars. We did this by multiplying each cost (capital or annual) by the ratio of the *average* Consumer Price Index values ("All Urban Consumers") for 1987 and 1981, respectively. (Admittedly, this escalation is an approximation, but as the 1981 costs were estimates already, little harm was done to the numbers.)

What do these expenditures tell us? First, in 1987 dollars, it cost a total of $25.6 billion to meet federally mandated air pollution control regulations in 1981, and would cost $320 billion to do so from 1981-90. Total capital costs for 1981 and 1981-90 were $10.5 and $128 billion, respectively. The "Mobile Sources" category comprises the largest share of the 1981 and 1981-90 capital costs—55% and 58%, in turn. Its share of the annual costs, however, is smaller than that contributed by the "Stationary Sources" categories—67% and 66% of the 1981 and 1981-90 totals.

This reversal is not too surprising if you consider that most of the Mobile Sources capital cost is for catalytic converters, modified ignition systems, and other control measures for automobiles and light-duty trucks. These devices are very capital-intensive, but require relatively little maintenance once installed. The only important annual costs are the capital charges and such O&M costs as fuel economy penalties and higher fuel costs (i.e., for unleaded gasoline). In the long run, however, Mobile Sources O&M costs are insignificant when compared to the capital charges. Although O&M comprises 26% of the 1981 Mobile Sources annual cost, it contributes only 4% to the 1981-90 annual cost.[8]

But with Stationary Sources, it's a much different story. Here, O&M

expenditures contribute much more to the annual costs than they do with Mobile Sources. To verify this, just compare their annual/capital cost ratios. For Mobile, the ratio is less than 1.5, while for Stationary it's nearly 4.0!

As Table 1.1 shows, "Electric Utilities" clearly dominates the Stationary Sources capital and annual costs both in 1981 and 1981-90. This again is hardly unexpected, given the immense expenditures incurred by utilities for emissions control (primarily particulate matter and sulfur oxides). Because they were compiled in 1981, the utility costs do not include any costs for acid rain prevention. Such costs are expected to significantly increase utilities' share of the cost pie. Although the "Governmental Expenditures" category accounts for only a small portion of the annual costs and none of the capital, it deserves special mention. This category includes ". . .estimates of expenditures made by federal, State, and local governments for both air pollution control programs and control of air pollution emitted from government facilities."[9] Thus, governments also bear a share of the cost for cleaning up the air. Among the federal agencies, the biggest spenders were the Departments of Defense and Energy and the Tennessee Valley Authority. The federal share of the total (including administrative, enforcement, and R&D costs) was 62% in 1981 and a projected 60% in 1981-90. State and local governments accounted for the rest of these expenditures.

But keep in mind that this report, though issued in 1984, was based on 1981 data. Had it been revised in a later year we might have seen considerably different results. The next two reports we'll encounter contain more recent cost data.

BUSINESSES AND AIR POLLUTION CONTROL

Once a year the Department of Commerce's Bureau of Economic Analysis (BEA) issues a short report with a long title: "Plant and Equipment Expenditures by Businesses for Pollution Abatement, 19--." This report presents estimates of the capital costs incurred by U.S. nonfarm businesses in year 19-- for new plants and equipment to control air and water pollution and dispose of solid waste. The report makes no distinction between pollution abatement equipment installed to comply with federally mandated regulations and that installed for other reasons. Thus, it provides estimates of *all* control investments made for whatever reason.[10]

The costs in the BEA report are "universe" estimates based on a survey of approximately 8000 companies, each of which is assigned to one of 24 industries corresponding to the industry classification for that company's principle product. BEA takes this information from its annual plant and equipment survey. These surveys contain capital cost data for both pollution control equipment (air, water, and solid waste) and total equipment. The pollution control cost data are grouped by industry and company size. Next, sample *ratios* of control costs to total plant and equipment spending are developed for

Table 1.2   Summary of BEA Report Air Pollution Control Capital Cost Estimates (million $), 1986-87

Industry Category	1986			1987		
	Total[a]	Air	%	Total	Air	%
Manufacturing						
Durable goods	71,660	1,110	1.5	70,670	860	1.2
Nondurable goods	76,240	1,380	1.8	74,310	1,360	1.8
Subtotal:	147,900	2,490	1.7	144,980	2,210	1.5
Nonmanufacturing						
Mining	11,630	80	0.7	10,100	40	0.4
Transportation	19,490	30	0.2	19,070	20	0.1
Public utilities	48,070	1,540	3.2	46,270	1,040	2.2
Trade and services	161,950	80	<0.1	167,740	80	<0.1
Communication and other	53,770	20	<0.1	53,810	10	<0.1
Subtotal:	294,910	1,750	0.6	296,980	1,190	0.4
TOTALS:[b]	443,000	4,240	1.0	442,000	3,410	0.8

Source: Rutledge, G. L., and N. S. Stergioulas. "Plant and Equipment Expenditures by Business for Pollution Abatement, 1986 and 1987," *Survey of Current Business*, October 1987, pp. 23-26. (All costs shown have been escalated to 1987 dollars.)
[a]"Total" refers to estimated total investment for the category in question, while "Air" denotes the investment for air pollution control only. Finally, "%" is the "Air"/"Total" ratio × 100%.
[b]Totals are rounded to three places.

each pollution control equipment category (e.g., air pollution stack controls) and industry size group. These sample ratios are then multiplied by *published* industry totals for all equipment spending to obtain estimates of total pollution control investment.[11]

To illustrate, consider the BEA air pollution control cost estimates for 1986 and the preliminary ("planned") data for 1987. Table 1.2 shows these data, along with estimates of total spending by industry category. As before, we have converted all estimates to constant (1987) dollars via the Consumer Price Index (CPI). Overall, total investments for air pollution control "plant and equipment" ranged from 0.8 to 1.0% of total spending for 1986–87. In all three years, over half of this cost was attributable to the "Manufacturing" subcategories, "Durable" and "Nondurable" goods. ("Durable goods" include such industries as primary metals; fabricated metals; machinery [electrical and other]; transportation; stone, clay, and glass; and other durables [lumber, furniture, instruments, and miscellaneous]. The "Nondurable goods" subcategory encompasses the following industries: food [including beverage], textiles, papers, chemicals, petroleum, rubber, and other nondurables [apparel, tobacco, leather, and printing-publishing].) No single industry dominated either subcategory's capital costs in these three years, however.

For the "Nonmanufacturing" category, however, public utilities alone comprised 35% of the all-industry total for air pollution controls. (Note that this is much less than the utilities' share of the total capital cost in the COCAW — 66% in 1981, for example.) As before, this is not surprising, given the large investments utilities have made in the past decade for emissions control. By comparison, the contributions made by other industries were insignificant.

Despite the billion-dollar magnitude of air pollution control capital costs, they still comprised only about 1% of total industry investments in 1986-87. Note that the costs shown in the BEA report comprised, overall, only about *one-half* of total business spending for pollution control (air, water, and solid waste). The rest was spent for such items as emission control devices on cars and trucks and agricultural and residential systems (e.g., septic systems).[12]

Can we meaningfully compare the COCAW and BEA report results? Probably not, if the above discrepancy in electric utility capital cost shares is any indication. For there are at least three areas in which the two reports differ: (1) scope of coverage, (2) types of costs reported, and (3) estimating procedure. First, as we've already seen, the COCAW covers a broad range of source categories — stationary and mobile — while the BEA report includes only the industry portion of the stationary sources sector. While the BEA reports only capital costs for plant and equipment, the COCAW reports both capital *and* annual costs for *all* types of control measures, not just those for plant and equipment. Finally, the BEA report is survey-based, while the COCAW is "computer model–based." Does this mean that the BEA report results are more credible because they're based on a statistically significant number of samples, rather than a computer model? Or is the COCAW more reliable because its scope is much broader than the BEA report? The answer to both questions is "No," because each report — and the Bureau of Census report we'll cover next — has advantages over the others. Thus, we would be foolish to select one report while rejecting the others. Instead, as we'll show later in this chapter, it's much better to combine the best features of two or more of the reports to produce a set of estimates superior to those in any of the individual publications.

CENSUS SURVEY

Annually, the Bureau of the Census surveys about 20,000 manufacturing establishments, each with 20 or more employees, regarding what they have spent in a given year to control air, water, or solid waste pollution.[13] From the data provided by the survey respondents, Census estimates the pollution control costs incurred by 19 manufacturing industry categories.

At first glance, it would seem that one part of the Commerce Department is duplicating the work of another. But such is not the case. Recall that the BEA survey covered a range of industrial *firms*, both manufacturing and nonmanufacturing, and reported for these firms air pollution control capital costs only. The Census report, on the other hand, presents both capital and annual costs, but only for *manufacturing establishments*, not firms.

These annual costs are comparable to the annual costs reported in the *Cost of Clean Air and Water* in that they include both depreciation and the traditional O&M costs—labor, materials and supplies, services, equipment leasing, and other expenditures. Although the Census report does not show itemized annual costs for air pollution control, it does break down the total for all pollutants (air, water, and solid waste). This breakdown is: depreciation (19.1%); labor (20.7%); materials and supplies (35.2%); and services, equipment leasing, and other costs (25.0%). Furthermore, this total annual cost excludes payments to governmental units for sewage treatment and other pollution control services. These payments amounted to approximately $10.7 billion in 1985.[14] However, these payments rarely apply to air pollution control, so whether they are included in total annual costs is immaterial.

Also reported with the annual costs were figures for "cost recovered." This item included the value of energy and materials recovered by control measures. For all forms of pollutants, "cost recovered" amounted to 11% of the 1985 operating cost.[15] Although the Census report did not do so, one could deduct this quantity from the annual cost to obtain a new measure—"net annual cost."

Although the annual costs were not itemized, the capital costs were, and quite extensively. For 21 major manufacturing categories (each of which consisted of two or more subcategories) air control capital costs were reported by abatement technique ("end-of-line" or "changes in production processes") *or* by type of pollutant abated. The latter included particulates, sulfur oxides, nitrogen oxides, carbon monoxide, hydrocarbon/volatile organic compounds, lead, hazardous air pollutants, and other.[16]

Finally, both capital and annual costs were shown by geographic region and state. For example, in 1985, three states contributed 28% of the total air control costs: California, Michigan, and Texas, in that order.[17] Moreover, Texas, along with California and Pennsylvania, accounted for 32% of the 1985 air control annual cost.[18] Keep in mind, though, that these costs are for *manufacturing* industries only. The Census report does not cover such large-emission sources as steam-electric power plants.

Although we don't have the space to summarize all the data in the Census report, we can show some of the highlights. Tables 1.3 and 1.4 display the capital and annual costs for the 19 industries surveyed by Census. Table 1.3 shows these costs for 1985, updated to 1987 dollars. Note that air pollution control expenditures account for nearly half (46%) of the capital costs spent to abate pollution from all three media—air, water, and solid waste. The "air share" of the annual costs is only 37%, however. Why? Mainly because, relative to their capital costs, water and solid waste control technologies have considerably higher annual costs—mainly operating and maintenance expenses. Thus, their share of the annual costs is higher. Moreover, as we said above, water and solid waste annual control costs also include costs for "user fees," while the air control costs do not include them.

That aside, let's examine the 1985 air control costs for a moment. In particular, note the annual/capital cost ratio: approximately 3. That ratio would be unusually high for an *individual* air pollution control system. But remember

Table 1.3 Summary of 1985 Census Report Control Costs

Industry Category	Capital (million $)			Annual (million $)		
	Total[a]	Air	%	Total	Air	%
Food	164	70	43	879	112	13
Tobacco	b	b	N.A.	37	17	46
Textile	26	13	50	157	33	21
Lumber	37	16	43	147	44	30
Furniture	16	10	63	63	25	40
Paper	340	202	59	1,184	331	28
Printing	42	32	76	125	55	44
Chemicals	780	205	26	2,684	711	26
Petroleum	306	185	60	2,194	1,351	62
Rubber	32	22	69	204	50	25
Leather	1	<0.5	N.A.	22	2	10
Stone	66	46	70	415	229	55
Primary metals	267	151	57	1,968	1,127	57
Fabricated metals	124	42	34	431	83	19
Machinery	73	22	30	363	80	22
Electric	146	48	33	452	82	18
Transportation	483	269	56	781	206	26
Instruments	26	15	58	174	25	14
Miscellaneous	b	b	N.A.	70	10	14
TOTALS:[c]	2,970	1,360	46	12,300	4,570	37

Source: Current Industrial Reports: Pollution Abatement Costs and Expenditures, 1985. Washington, DC: U.S. Department of Commerce, Bureau of the Census (Report No. MA-200(85)-1), April 1987. (All costs shown have been escalated to 1987 dollars.)
[a]"Total" refers to total pollution control cost for the category, while "Air" denotes the air pollution control cost only. Finally, "%" is the "Air"/"Total" ratio × 100%.
[b]Indicates data withheld to avoid disclosing operations of individual companies.
[c] Totals are rounded to three places.

that these annual cost figures represent not only those for systems brought on-line in 1985, but also those started up in *previous* years. In this respect, the Census report is like the COCAW. Scanning down the columns, however, we see that the annual/capital cost ratio varies by industry category, and noticeably so. For "Food," for example, the ratio is about 1.5, while for "Petroleum" it's over 7. Does this mean that vastly different kinds of control devices are used in the food and petroleum industries? Probably not. What it *likely* means, though, is that the petroleum industry has been controlling air pollution for a longer time than has the food category. That is, the petroleum industry has a lot more equipment in place that it has to keep paying for to operate, maintain, and depreciate.

Which of these categories are the biggest spenders? In 1985, "Petroleum" topped the annual cost list, with "Primary Metals" a close second, followed by "Chemicals" and "Paper." Because the annual cost includes depreciation — and thereby reflects the 1985 capital cost — it is a better measure of control expenditure than the capital cost. Nonetheless, for the record, the leading categories in terms of 1985 investment were "Transportation," "Chemicals," "Paper," and "Petroleum." (*Note:* "Transportation" includes those industries that manufac-

Table 1.4   Summary of 1986 Census Report Control Costs

Industry Category	Capital (million $)			Annual (million $)		
	Total[a]	Air	%	Total	Air	%
Food	193	64	33	967	132	14
Tobacco	b	b	N.A.	35	15	43
Textile	26	12	46	168	30	18
Lumber	34	19	56	187	64	34
Furniture	22	16	73	102	31	30
Paper	281	142	51	1,197	331	28
Printing	26	19	73	151	61	40
Chemicals	647	205	32	2,752	671	24
Petroleum	439	283	64	2,078	1,276	61
Rubber	37	21	57	234	53	23
Leather	3	1	33	31	2	6
Stone	87	55	63	439	246	56
Primary metals	234	107	46	1,805	1,003	56
Fabricated metals	141	38	27	521	100	19
Machinery	51	18	35	371	84	23
Electric	130	49	38	552	91	16
Transportation	561	448	80	889	205	23
Instruments	20	11	55	170	20	12
Miscellaneous	b	b	N.A.	88	6	7
TOTALS:[c]	2,950	1,520	52	12,700	4,420	35

Source: Current Industrial Reports: Manufacturers' Pollution Abatement Capital Expenditures and Operating Costs--Advance Report for 1986. Washington, DC: Bureau of the Census (Report No. MA-200(86)-1), December 1987. (All costs shown have been escalated to 1987 dollars.)
[a]"Total" refers to total pollution control cost for the category, while "Air" denotes the air pollution control cost only. Finally, "%" is the "Air"/"Total" ratio × 100%.
[b]Indicates data withheld to avoid disclosing operations of individual companies.
[c]Totals are rounded to three places.

ture motor vehicles, aircraft, ships, and other transportation equipment. It does *not* include the costs of controlling emissions from that equipment.)

The 1986 results are similar to those for 1985. However, these data were taken from the Census Bureau's "Advance Report—1986," which did not provide the breadth and depth of data that the full 1985 report supplied.[19] Again, "Petroleum," "Primary Metals," and "Chemicals" lead the annual cost list, while "Transportation," "Petroleum," and "Chemicals" were the top contributors to the capital cost. Overall, the air control capital cost increased by about 12% from 1985 to 1986, while the annual cost *decreased* by a much smaller amount, 3%.

It's also interesting to compare the annual/capital cost ratios for these two years. The composite ratio—2.9—is about the same as the 1985 ratio. But, for some of the smaller categories the ratio changed dramatically. For "Food," for example, the ratio grew from 1.6 to 2.1, while "Lumber's" ratio increased from 2.8 to 3.4. But for the larger categories, the ratio stayed about the same. What does this indicate? Possibly that the larger categories *replaced* a lot of their existing controls with replacement equipment in 1986. This, in turn, meant that the annual costs incurred by the equipment in past years were "substituted for" by annual costs for the replacement equipment. Therefore, any growth in the annual costs for these larger categories between 1985 and 1986 was solely due to the relatively small cost additions contributed by the replacement control systems. Thus, the annual/capital cost ratios stayed about the same for those industries with large amounts of equipment to replace. However, for those industries with little or no equipment to replace (e.g., "Food"), the annual cost grew at a faster rate than the capital cost, because the depreciation and O&M costs for the existing equipment had to be "carried along" from 1985 to 1986. Consequently, for these industries, the annual/capital cost ratios grew.

Finally, let's stand back from the numbers in Tables 1.3 and 1.4 and ask ourselves: "Just how *accurate* are these costs anyway?" Anticipating questions like that, the Census Bureau has calculated the "standard error" of each estimate reported. For the 19 individual categories, this statistic ranges from 1 to 42%, with the smaller errors generally associated with those industries having reported the larger costs and vice versa. More importantly, the standard error of the *total* capital and annual costs is a mere *1%,* for both 1985 and 1986.

Of the three reports we've examined in this chapter, the Census study is the only one that provides explicit accuracy information. This allows us to place confidence limits around the reported costs. Moreover, the fact that the standard error is only 1% for the total and many of the individual costs lends a great deal of credibility to the Census data. But despite the high quality of the Census estimates, they are reported for *manufacturing* industries only. Is there a way we can somehow extrapolate the Census results to obtain cost estimates for a full range of air pollution source categories? And can we use these extrapolated data to make future cost projections? We'll find out in the next section.

"HYBRID" NATIONAL CONTROL COST ESTIMATES

Of the three reports we've discussed, only the *Cost of Clean Air and Water* contains control costs for most source categories. Unfortunately, the latest edition of the report contains seven-year-old data which are of unknown (and probably uneven) accuracy. Given these limitations, how can we make use of the COCAW data to develop more recent national control cost estimates? First, we must realize that the COCAW *absolute* costs are, due to their age, of limited value. However, the *relative* costs are quite useful, for we can reasonably expect them to remain constant (or nearly so) from year to year. Thus, if we somehow "married" these relative costs with *current* absolute costs obtained from another reference, we could obtain fairly accurate costs for a wide range of source categories.

This is, in fact, what we'll do. First, we have to decide which of the two Department of Commerce reports offered the most complete data. Although the BEA report provided costs for a wider range of stationary sources than the

Census report, it only reported capital costs, while the Census publication showed both capital and annual expenditures. Therefore, we'll choose the Census report data as the foundation of our hybrid estimates.

Our next step is to identify those source categories in the two reports which correspond (or approximate a correspondence). Of the stationary source categories listed in Table 1.1 for the COCAW, four correspond to the industrial categories in the Census report: "Chemicals" ("Chemicals" in the Census report), "Food Processing" ("Food"), "Metals" ("Primary Metals"), and "Softgoods" ("Paper"). There may also be full or partial correspondence among some of the other categories, but the information available in the COCAW and Census reports does not permit us to fully determine it.

This correspondence established, the next steps in calculating the hybrid (extrapolated) costs are:

1. *Sum* the 1981 capital and annual costs in Table 1.1 for the "Food Processing," "Chemicals," "Metals," and "Softgoods" source categories. These sums are $576 million (capital) and $3,770 million (annual).

2. *Divide* these sums into the 1981 capital and annual costs in Table 1.1 for the other stationary sources categories and for the "Mobile Sources" and "Government Expenditures" categories. These quotients are the capital and annual cost "extrapolation ratios" for these categories. (Example: "Electric Utilities" capital cost ratio = $3,120/$576 = 5.42.)

3. *Sum* the 1985 and 1986 capital and annual costs in Tables 1.3 and 1.4, respectively, for the "Food," "Paper," "Chemicals," and "Primary Metals" categories. These sums appear in Tables 1.5 (1985) and 1.6 (1986).

4. *Multiply* the sums from Step 3 by the extrapolation ratios calculated in Step 2 to obtain the extrapolated costs. (Example: for "Electric Utilities" in 1986, capital cost [millions] = $1,360 × 5.42 = $7,370.)

Tables 1.5 and 1.6 show the results of this procedure. The extrapolated air control costs (indicated by the footnotes) are, of course, approximations, based on the above ratios and the category correspondence established between the two reports. Nonetheless, the results are interesting. The estimated total capital costs are roughly $30 and $26 billion in '85 and '86, respectively. The total annual costs for these two years are lower—$18 and $17 billion, in turn.

In both 1985 and 1986, the "Mobile Sources" category dominates the capital cost, accounting for over half the total in both years. "Electric Utilities" is the single biggest contributor to the annual costs in both years. This "flip-flop" is due to the different annual/capital cost ratios for these two categories. As stated earlier, mobile sources controls incur relatively low O&M costs, com-

Table 1.5   National Air Pollution Control Cost Estimates—1985

Source Category	Capital (million $)	Extrap. Ratio	Annual (million $)	Extrap. Ratio
Census Report				
Food, Paper, Chemicals, and Primary Metals	1,550	N.A.	2,280	N.A.
Other Census categories	1,420	N.A.	2,290	N.A.
Subtotal:	2,970	N.A.	4,570	N.A.
COCAW Report [a]				
Electric utilities	8,400	5.42	5,630	2.47
Other fuels and energy	403	0.260	145	0.0637
Construction materials	161	0.104	224	0.0981
Manufacturing	403	0.260	91	0.0398
Municipal waste incineration	27	0.0174	18	0.00796
Other industrial	1,800	1.16	2,050	0.897
Mobile sources	15,420	9.95	4,570	2.01
Government expenditures	0	0	461	0.202
Subtotal:	26,610	N.A.	13,190	N.A.
TOTALS:[b]	29,600	N.A.	17,800	N.A.

Source: See Tables 1.1, 1.3, and 1.4. (All costs shown have been escalated to 1987 dollars.)
[a]COCAW report costs have been extrapolated from Census report costs.
[b]Totals have been rounded to three places.

pared to the flue gas desulfurization systems and other controls on power plants. Apart from these two, however, no other source categories stand out, indicating a fairly even distribution of the costs.

Now that we have extrapolated from the Census report data to obtain costs

Table 1.6   National Air Pollution Control Cost Estimates—1986

Source Category	Capital (million $)	Extrap. Ratio	Annual (million $)	Extrap. Ratio
Census Report				
Food, Paper, Chemicals, and Primary Metals	1,360	N.A.	2,140	N.A.
Other Census categories	1,590	N.A.	2,280	N.A.
Subtotal:	2,950	N.A.	4,420	N.A.
COCAW Report [a]				
Electric utilities	7,370	5.42	5,290	2.47
Other fuels and energy	354	0.260	136	0.0637
Construction materials	141	0.104	210	0.0981
Manufacturing	352	0.260	85	0.0398
Municipal waste incineration	24	0.0174	17	0.00796
Other industrial	1,570	1.16	1,920	0.897
Mobile sources	13,480	9.95	4,300	2.01
Government expenditures	0	0	432	0.202
Subtotal:	23,270	N.A.	12,390	N.A.
TOTALS:[b]	26,200	N.A.	16,800	N.A.

Source: See Tables 1.1, 1.3, and 1.4. (All costs shown have been escalated to 1987 dollars.)
[a]COCAW report costs have been extrapolated from Census report costs.
[b]Totals have been rounded to three places.

Table 1.7   Summary of Census Report Air Pollution Control Costs: 1980-86

Year	Cost (millions of 1987 $)[a]	
	Capital	Annual
1980	2,390	4,580
1981	2,740	4,620
1982	2,150	4,070
1983	1,170	4,340
1984	1,140	4,580
1985	1,370	4,570
1986	1,520	4,420

Source: *Current Industrial Reports: Pollution Abatement Costs and Expenditures—1980, 1985, 1986 (Advance Report)*. Washington, DC: U.S. Department of Commerce, Bureau of the Census.
[a]All costs have been rounded to three places.

for the COCAW source categories, can we now *project* these extrapolated figures to future years? Before answering that question, we need to examine historical air pollution costs. Again, the Census report is the best source of information. As Table 1.7 shows, neither the capital nor the annual costs reported by Census have exhibited any clear trends since 1980.[20,21] The capital costs decreased from a 1981 high to a 1984 low and then took an upturn. Meanwhile, the annual costs fluctuated almost sinusoidally from year to year, although the amplitude of the curve was relatively small, the difference (in 1987 dollars) between the lowest and highest annual costs being only 12%. For these reasons, it would be risky to attempt to project future costs from these data. Moreover, unlike the cost of living, which *always* increases, there is no evidence to show that control costs will keep growing indefinitely.

*   *   *

So much for the "bottom line" — or bottom *lines*, if you will. Large numbers like the foregoing let us view air pollution control in "macroeconomic" terms. Some may use these costs to argue that we are already spending too much for air pollution control and that we should relax regulations at *all* levels of government. Others, however, upon comparing these billion-dollar expenditures to the trillion-dollar federal budgets and multi-trillion dollar Gross National Products, might conclude that, relatively speaking, air pollution control is a bargain. Whichever viewpoint one embraces, the control costs are *here*, to be used to argue for or against more stringent regulation.

The rest of this book deals with how to arrive at estimates of the costs which, when aggregated, would produce the "bottom line" costs shown in this chapter. Using a simple analogy, the aggregated costs are the building, while the costs of individual control technologies are the bricks and lumber, and the costing procedures and guidelines are the mortar and nails holding everything together. If you're satisfied just to look at the building, *read no further*. You've probably learned all you want to know about air pollution control costs. If, however, you're interested in how to draw the plans, select the building materials, and erect the structure, *the rest of this book is for you.*

REFERENCES

1. *The Cost of Clean Air and Water: Report to Congress.* Washington, DC: U.S. Environmental Protection Agency, Office of Policy Analysis, May 1984 (hereinafter cited as *Cost of Clean*), pp. 9-10.
2. Clean Air Act, as amended (through July 1981). Serial No. 97-4, U.S. Government Printing Office, Washington, DC, 1981.
3. *Cost of Clean*, p. 1.
4. *Cost of Clean*, pp. 18-20.
5. *Cost of Clean*, pp. 18-20.
6. *Cost of Clean*, p. 19.
7. *Cost of Clean*, p. 18.
8. *Cost of Clean*, p. A4-19.
9. *Cost of Clean*, p. A2-1.
10. Rutledge, G. L., and N. S. Stergioulas. "Plant and Equipment Expenditures by Business for Pollution Abatement, 1986 and 1987," *Survey of Current Business,* October 1987 (hereinafter cited as *BEA Survey, 1986-87*), pp. 23-26.
11. "Plant and Equipment Expenditures by Business for Pollution Abatement: Revised Estimates for 1973-83; Estimates for 1984," *Survey of Current Business,* February 1986, pp. 39-45.
12. *BEA Survey, 1986-87.*
13. *Current Industrial Reports: Pollution Abatement Costs and Expenditures, 1985.* Washington, DC: U.S. Department of Commerce, Bureau of the Census (Report No. MA200[85]-1), April 1987 (hereinafter cited as *Census Report, 1985*), p. x.
14. *Census Report, 1985*, p. v.
15. *Census Report, 1985*, p. 44.
16. *Census Report, 1985*, p. 8.
17. *Census Report, 1985*, p. 18.
18. *Census Report, 1985*, p. 36.
19. *Current Industrial Reports: Manufacturer's Pollution Abatement Capital Expenditures and Operating Costs — Advance Report for 1986.* Washington, DC: U.S. Department of Commerce, Bureau of the Census (Report No. MA200-[86]-1), December 1987.
20. *Current Industrial Reports: Pollution Abatement Costs and Expenditures, 1980.* Washington, DC: U.S. Department of Commerce, Bureau of the Census (Report No. MA-200[80]-1), December 1981, p. 12.
21. *Census Report, 1985*, p. 1.

CHAPTER 2

Exploring Cost Engineering

Who shall doubt "the secret hid
Under Cheops' pyramid"
Was that the contractor did
Cheops out of several millions?

A General Summary — Rudyard Kipling

Vince Lombardi, the legendary Green Bay Packers coach, believed that every player, regardless of his skills and accomplishments, should have a thorough knowledge of football fundamentals. Accordingly, on the first day of preseason training camp, he would gather his players around him, rookies and seasoned veterans alike. While holding up the old pigskin, Lombardi would begin his lecture on the fundamentals: "Now men, *this* is a football."

The fundamentals are important to every field of endeavor, and cost engineering is no exception. Several books written under the auspices of the American Association of Cost Engineers (AACE) present these fundamentals quite well.[1-3] So do such classics as *Engineering Economy* by Grant, Ireson, and Leavenworth.[4] It wouldn't do to duplicate these works here. Instead, in this chapter we'll present the basics of cost engineering, focusing mainly on the concepts that apply to the costing of air pollution control techniques.

COST TERMINOLOGY

Cost engineering would be a far easier discipline to master if estimators used a consistent set of terminology. For instance, the following terms are among those used to denote the costs incurred yearly during the term (or "useful life") of a project: "annualized cost," "operating and maintenance (or O&M) cost," "total annual cost," and "net cash flow." All supposedly mean the same thing. Unfortunately, some estimators interpret these quantities to mean different things. To some, for example, "O&M" costs mean just the portion of the total annual cost attributable to labor, maintenance, and utilities expenses. To avoid this confusion and to maintain consistency, we'll follow the AACE terminology as much as possible in this book.

17

Total Capital Investment

With "add-on" and most other air pollution control systems, a lump sum outlay must be made before the system can be built or placed in operation. This "capital" expenditure covers the cost of purchasing and installing the control devices and auxiliary equipment, as well the cost of land, buildings, and "offsite" facilities. Even with those systems where no new equipment is needed, there may still be costs for supplies of operating materials (such as low-solvent coatings) that are needed before the system is started up. These costs and others make up the *total capital investment* (TCI). The TCI is comprised, in turn, of "depreciable" and "nondepreciable" investments. By "depreciable" we refer to those costs that can be depreciated during all or part of the time the control technology is in operation (i.e., its useful life). Depreciable costs include the above equipment and facilities-related costs plus any others that can be "written off" on income tax returns. (We'll discuss depreciation more thoroughly later in this chapter.)

Conversely, such costs as land and working capital comprise the *nondepreciable* elements of the TCI, because they are said to be "recoverable" when the control technology ceases operation. Moreover, income tax regulations do not permit these costs to be depreciated on tax returns.

Table 2.1 lists the many items comprising the total capital investment. Note that the total depreciable investment has two components: the "battery limits" and "offsite facilities" costs. The latter consists of those facilities that, though not an integral part of the control system, are needed to support its operation. (An electrical substation erected to supply power for an energy-intensive control system would be an example.) "Add-on" control systems typically require no offsite facilities, but others, such as flue gas desulfurization systems (FGDs) typically do.

The "battery limits" cost is broken down into "direct" and "indirect" costs. *Direct* costs cover the cost of the equipment (including freight and applicable sales taxes), such "hard" installation costs as piping and electrical work, site preparation, and buildings. The *indirect* ("soft") installation costs comprise engineering costs, construction and field expenses (e.g., rental of trailers and like equipment), contractor fees (for firms involved in the project), startup and performance tests (to get the control system running and to verify that it meets the vendor's guarantees), and contingencies.

"Contingencies" is more an afterthought item than a true cost. It simply covers all those expenses that the estimator cannot foresee, but which could very well be incurred. Examples are drastic increases in equipment prices, construction labor shortages or stoppages, delays of all kinds, or damage due to violent weather.

With some estimating procedures, the direct and indirect installation costs are "factored" from the equipment cost. (That is, the purchased cost is multiplied by certain percentages to compute the individual installation costs.) Installation factors for one class of controls ("add-ons") are shown in Table

Table 2.1   Components of Total Capital Investment (TCI)

TCI = Depreciable Investment + Nondepreciable Investment

"Nondepreciable Investment" consists of land and working capital.

"Depreciable Investment" consists of:
Offsite facilities
Total Indirect Cost[a]
 − Indirect Installation Costs
   • Engineering and supervision
   • Construction and field expenses
   • Construction fee
   • Startup
   • Performance test
   • Contingencies

Total Direct Cost[a]
 − Site Preparation
 − Buildings
 − Purchased Equipment Cost
   • Control device
   • Auxiliary equipment
   • Instrumentation
   • Sales taxes
   • Freight

 − Direct Installation Cost
   • Foundations and supports
   • Handling and erection
   • Electrical
   • Piping
   • Insulation
   • Painting

[a]The sum of the "Total Direct" and "Total Indirect" costs is sometimes termed the "Battery Limits" cost.

2.2. These factors represent typical installation conditions. Where these conditions differ, however, the installation factors could be much higher or lower. To compensate for these differences, "adjustment multipliers" are applied against the "base" installation factors to obtain the installation factors for the situation in question. Both the factors and multipliers were obtained from reference 5. (Example: for a hazardous process, requiring a safety backup, the adjustment multiplier for "instrumentation" in Table 2.3 is 3. From Table 2.2, the base instrumentation factor is 0.10. Therefore, the *adjusted* instrumentation factor would be $3 \times 0.10 = 0.30$.) These factors and multipliers would only be used with factored-type estimates. With other types of estimating methods, other types of factors would be used. Finally, in some cases, the various installation costs are calculated individually. (Some of the different estimating methods will be discussed later in this chapter.)

As discussed above, the *nondepreciable* costs include "land" and "working capital." The amount of land needed with a control system will vary considerably. "Add-ons" may require no more than a quarter-acre, while an FGD may need several acres for the waste treatment pond alone. The cost of land also

Table 2.2 Average Cost Factors for Selected "Add-on" Control Devices

Cost Item	Electrostatic Precipitator	Venturi Scrubber	Fabric Filter
Direct Costs			
• Purchased equipment (PE)			
Control device	X		
Auxiliaries	Y		
Instrumentation	0.10 (X + Y)	(Same factors for all "add-ons")	
Taxes	0.03 (X + Y)		
Freight	0.05 (X + Y)		
Total PE = Z =	1.18 (X + Y)		
• Installation			
Foundations, supports	0.04Z[a]	0.06Z	0.04Z
Handling, erection	0.50	0.40	0.50
Electrical	0.08	0.01	0.08
Piping	0.01	0.05	0.01
Insulation	0.02	0.03	0.07
Painting	0.02	0.01	0.02
Total direct installation	0.67Z	0.56Z	0.72Z
• Site preparation	———— As required ————		
• Facilities, buildings	———— As required ————		
Indirect Costs			
• Installation			
Engineering, supervision	0.20Z	0.10Z	0.10Z
Construction, field	0.20	0.10	0.20
Construction fee	0.10	0.10	0.10
Startup	0.01	0.01	0.01
Performance test	0.01	0.01	0.01
Model study	0.02	–	–
Contingencies	0.03	0.03	0.03
Total indirect installation	0.57Z	0.35Z	0.45Z

varies, depending upon the location of the facility, local zoning and building codes, and other considerations.

The kinds and amounts of working capital needed are also quite variable. Some systems (e.g., fabric filters) require none, while others may need a great deal (e.g., an oil-fired incinerator, which could require a 30-day initial supply of fuel). Working capital is normally incurred with larger, more complex control technologies – specifically, those that require significant stockpiles of raw materials, utilities, and other consumables on hand before startup. Flue gas desulfurization systems are illustrative. According to Uhl,[6] working capital (when applicable) may be estimated at 10 to 15% of the TCI or 15 to 35% of the revenue. But because control technologies rarely generate revenues, we can replace "revenue" with "direct annual cost."

Another expenditure incurred by larger projects is "interest during construction" (IDC). IDC covers the cost of borrowing funds for making installment payments to contractors while the control system is being built. These monies are repaid once the project is complete and capital is available from bond or stock issues. Estimating IDC requires knowledge of the borrowing rate and the

Table 2.2, continued

Cost Item	Thermal, Catalytic Incinerator	Carbon Adsorber	Gas Absorber Flare
Direct Costs			
• Purchased equipment (PE)			
Control device	X		
Auxiliaries	Y		
Instrumentation	0.10 (X + Y)	(Same factors for all "add-ons")	
Taxes	0.03 (X + Y)		
Freight	0.05 (X + Y)		
Total PE = Z =	1.18 (X + Y)		
• Installation:			
Foundations, supports	0.08Z[a]	0.08Z	0.12Z
Handling, erection	0.14	0.14	0.40
Electrical	0.04	0.04	0.01
Piping	0.02	0.02	0.30 (absorber) 0.02 (flare)
Insulation	0.01	0.01	0.01
Painting	0.01	0.01	0.01
Total direct installation	0.30Z	0.30Z	0.85Z (absorber) 0.57 (flare)
• Site preparation	———————— As required ————————		
• Facilities, building	———————— As required ————————		
Indirect Costs			
• Installation:			
Engineering, supervision	0.10Z	0.10Z	0.10Z
Construction, field	0.05	0.05	0.10
Construction fee	0.10	0.10	0.10
Startup	0.02	0.02	0.01
Performance test	0.01	0.01	0.01
Contingencies	0.03	0.03	0.03
Total indirect installation	0.31Z	0.31Z	0.35Z

Source: Vatavuk, W. M., and R. B. Neveril. "Estimating Costs of Air-Pollution Control Systems, Part II: Factors for Estimating Capital and Operating Costs," *Chemical Engineering*, November 3, 1980, pp. 157-162.
[a]Each of the direct and indirect installation factors is multiplied by "Z," the Purchased Equipment cost (PE).

loan and payment schedules. IDC is only relevant in cases where the construction period exceeds one year, however. Consequently, it is rarely included in control system capital costs, because most systems can be fabricated and installed in much less time. Flue gas desulfurization systems are an exception (again).

The land and working capital are not the sole "recoverable" costs, however. In some cases, all or part of the control equipment may be *salvageable*, thus offsetting the TCI. However, this offset is watered down by the fact that it occurs at the *end* of the system's useful life, while the other capital costs are

Table 2.3 Adjustment Multipliers for Add-On Installation Factors

Installation Cost	Adjustment Factor
Instrumentation	
• Simple, continuous, manual	0.5 to 1.0
• Intermittent control	1.0 to 1.5
• Hazardous process with safety backup	3
Freight	
• Major urban areas in continental U.S.	0.2 to 1.0
• Remote areas in continental U.S.	1.5
• Alaska, Hawaii, and other countries	2
Handling and Erection	
• Delivered cost includes assembly; supports, base, skids provided	0.2 to 0.5
• Modular equipment; compact size	1
• Fabrication onsite, with extensive welding and erection	1 to 1.5
• Retrofitting existing plant; includes equipment removal and site renovation	2
Site Preparation	
• Minimum clearing and grading	0
• Extensive leveling and removal of structures; includes land survey and study	1
• Extensive excavation, filling, and leveling; may include draining and pile-setting	2
Facilities and Buildings	
• Outdoor process; utilities at site	0
• Outdoor process with some weather enclosures; utilities brought to site; access roads, fencing, and minimum lighting	1
• Buildings with heating and cooling, sanitation facilities; shops and office; may include railroad sidings, truck depot, and parking	2
Engineering and Supervision	
• Standard equipment, duplicate of typical system; turnkey quotation	0.5
• Custom equipment; automatic control	1 to 2
• Prototype equipment	3
Construction and Field	
• Small-capacity system	0.5
• Medium-capacity system	1
• Large-capacity system	1.5
Construction Fee	
• Turnkey; erection and installation included in equipment cost	0.5
• Single contractor for installation	1
• Primary contractor and several subcontractors	2
Contingency	
• Developed process	1
• Prototype or experimental process	3 to 5
• Pilot tests required to obtain efficiencies and operating specifications guarantees	5 to 10

Source: Vatavuk, W. M., and Neveril, R. B. "Estimating Costs of Air-Pollution Control Systems, Part II: Factors for Estimating Capital and Operating Costs," *Chemical Engineering,* November 3, 1980, pp. 157-162.

incurred at the *beginning*. Thus, the "present worth" of the salvage is much less. (See discussion below on "measures of merit.") Secondly, if salvage is introduced, the amount of the TCI that can be depreciated must be reduced by the salvage value. With most "add-on" control systems, however, a salvage value is rarely taken, because the equipment is designed and sized for a particular emission source and cannot be used at another facility without making extensive modifications. Moreover, the cost of disassembling and removing the equipment often exceeds its salvage value.

Finally, not all capital costs are incurred at the beginning of a project. Some control systems require replacement of major components before the end of the system's useful life. (Example: replacement carbon for a carbon adsorber.) These expenditures must be treated as capital costs, not recurring (annual) costs.

Total Annual Cost

As the name indicates, *annual* costs are those expenses incurred every year during the life of a control system. Due to inflation and other factors, the *magnitude* of these annual costs may not be the same every year. However, the same *kinds* of costs will be incurred.

Using AACE terminology, the *total annual cost* (TAC) is comprised of two elements: the *direct* annual cost (DAC) and the *indirect* annual cost (IAC). These costs are offset by *recovery credits* (RC). These credits are due to energy and materials recovered by the control system, such as the perchloroethylene solvent captured by carbon adsorbers in drycleaning establishments and recycled to the cleaning apparatus. From a "cash flow" standpoint, recovery credits are considered revenue or income. (We'll discuss cash flow in the next section.)

The following equation ties together the three elements of the TAC:

$$TAC = DAC + IAC - RC \qquad (2.1)$$

where the units of each term are typically \$/year

Direct annual costs are ". . .those which tend to be proportional or partially proportional to throughput or production," while *indirect* annual costs are those ". . .which tend to be independent of production. These costs are incurred whether or not production rates change."[7] Direct costs are, in turn, subdivided into two categories: *variable* and *semivariable*. Variable costs are those expenses whose values are directly proportional to the plant production level or (usually in the case of control systems) the volume of waste gas processed by the system. Utilities (electricity, fuel, steam, water, etc.), raw materials, waste treatment/disposal, and any other items consumed in direct proportion to the production level/gas volume processed comprise the variable costs. In other words, when the control system is operating at maximum capacity, variable costs are highest. Conversely, when the control system is shut down, the variable costs are zero.

Semivariable costs also depend on the control system gas throughput, but only partially. Labor costs (operating, maintenance, supervisory), payroll overhead, maintenance materials, and recurring replacement parts are examples of semivariable costs. Even at zero gas throughput, semivariable costs are from 20 to 40% of their value at full throughput.[8]

By definition, *indirect* annual costs are completely independent of the gas throughput. One hundred percent of these expenditures would be incurred even if the control system were shut down. Also called *fixed* costs, indirect costs include plant overhead, property taxes, insurance, general and administrative expenses, and depreciation/capital recovery. (Note: there is a distinct difference between depreciation and capital recovery, which will be explained below.) The various annual costs are shown in Table 2.4.

Let's take a closer look at these annual costs.

Utilities

As stated above, utilities consist of electricity, fuel, steam, water (process or cooling), fuel, compressed air, and other items that control systems consume. Some utilities are usually purchased (e.g., electricity and fuel), while others (water, steam) are typically produced/processed onsite.

With some control systems, the amount of utilities consumed can be quite significant and can overshadow the other costs. A thermal incinerator is illustrative. Here, fuel can make up 75% or more of the system total annual cost, especially if the recuperative heat recovery is low. Conversely, with an electrostatic precipitator, utility consumption (electricity) is relatively low. Although the kinds and quantities of utilities consumed by control systems will vary, nearly all will require electricity for powering the system fan. The fan power requirement (in horsepower) can be calculated from the standard formula:

$$FP = 0.0001575\Delta PQ/n \qquad (2.2)$$

where ΔP = system pressure drop (inches of water, column)
Q = waste gas volumetric flowrate through system (actual cubic feet/minute)
n = combined efficiency of fan and motor (generally, 0.50 to 0.70)

Electricity is also used for powering pumps (scrubbers), rectifiers (electrostatic precipitators), and shaker motors (fabric filters), to name a few applications.

Other utilities and examples of control technologies that consume them are:

- steam: carbon adsorbers
- compressed air: fabric filters (pulse-jet type)
- refrigeration: refrigerated condensers

Table 2.4   Components of Total Annual Cost (TAC)

TAC = Direct Annual Costs + Indirect Annual Costs − Recovery Credits

Direct Annual Costs
Variable:
  − Raw materials
  − Utilities
  • Electricity
  • Steam
  • Fuel
  • Water
  • Compressed air
  • Others
  − Waste treatment/disposal

Semivariable:
  − Labor
  • Operating
  • Supervisory
  • Maintenance
    − Maintenance materials
    − Replacement parts
    − Payroll overhead

Indirect Annual Costs
Capital recovery/depreciation
Plant overhead
Property taxes
Insurance
Administrative charges

Recovery Credits
Materials
Energy

- fuel: incinerators (catalytic and thermal)
- water: wet dust suppression system

As we'll see in later chapters of this book, the quantities of utilities required are calculated based on the operating and design features of the control system. Therefore, with the exception of the fan power formula above, we'll not provide any specifics here on determining utility requirements. We can, however, list some sources of utility *price* data:

- electricity: Energy Information Administration (EIA), (U.S. Department of Energy)
- fuel: EIA, Electric Power Research Institute
- water: American Association of Waterworks

Raw Materials

The kinds and quantities of raw materials required depend on both the size and *type* of the control technology. Some, such as fabric filters, require no raw materials, while FGDs typically consume large amounts — and even require

Table 2.5 Operating and Maintenance Labor Factors for Selected "Add-on" Control Devices

	Labor Factor (hours/shift)	
Control Device	Operating	Maintenance
Fabric filter	2 to 4	1 to 2
Electrostatic precipitator	1/2 to 2	1/2 to 1
Venturi scrubber	2 to 8	1 to 2
Incinerator	1/2	1/2
Gas absorber	1/2	1/2
Carbon adsorber	1/2	1/2
Refrigerated condenser	1/2	1/2
Flare	–	1/2

Source: Vatavuk, W. M., and R. B. Neveril. "Estimating Costs of Air-Pollution Control Systems, Part II: Factors for Estimating Capital and Operating Costs," *Chemical Engineering*, November 3, 1980, pp. 157-162.

special equipment for preparation of these materials before they are used. In addition, some systems (e.g., wet scrubbers) may require chemicals for treatment of wastewater and sludges generated by the capture of air pollutants. Some other technologies and the raw materials they consume are:

- wet scrubbers: caustics, coagulants
- gas absorbers: absorbents
- wet suppression systems: surfactants
- low solvent coating processes: coatings

The amounts of materials needed are calculated from process requirements (e.g., estimated makeup rates), material balances, etc. Costs for the materials are available from suppliers or publications such as the *Chemical Marketing Reporter*.

Operating Labor

Nearly all control techniques require at least some operator attention. This can range from a few minutes per shift to the assignment of several full-time personnel. No precise factors can be given for estimating this labor requirement, as it will vary with the size, complexity, level of automation, and operating mode of the technique. However, some rough guidelines have been developed for estimating operating and maintenance labor requirements for "add-on" control systems.[9] (See Table 2.5.) These values pertain to relatively large, automated, continuously operated systems. The ranges shown in Table 2.5 reflect corresponding ranges in control system production level.

Other references have provided equations that display the labor requirement–production level relationship. One source[10] postulated a logarithmic relationship, viz.:

$$L = a(X)^b \tag{2.3}$$

where   L = labor requirement
         X = production level
      a, b = parameters

The difficulty with this and similar relationships is that it doesn't fit all situations. For instance, Equation 2.3 doesn't account for the fact that even the smallest system requires a minimum amount of operating (and maintenance) labor. That is, as X approaches zero, so does L. An improved form of this equation would include a constant term ("c"), which would allow for this minimum labor requirement. In any case, the estimator is best advised to rely on past experience when estimating labor requirements.

In addition to operating labor, one should add from 10 to 20% more to cover the cost of supervisory labor (i.e., foremen, shift supervisors). A reasonable average is 15%.[11] That is, if the operating labor requirement were 10 hours/shift, the supervisory labor required would be 1.5 hours/shift. Typically, the supervisory labor *rate* ($/hour) would be different from the operating labor rate. Values for labor rates can be found in such publications as the *Monthly Labor Review* (U.S. Department of Labor, Bureau of Labor Statistics) and in similar reports issued by state labor departments. Headquarters of international unions (e.g., Oil, Chemical, and Atomic Workers) are also sources of wage data.

Maintenance

In most cases, maintenance cost is estimated by one of two different methods. In the first method, the total capital investment (or sometimes just the total direct cost) is multiplied by some factor, usually 2 to 5%. The factor varies with the type of the control technique, and is usually chosen subjectively by the estimator. The advantage of this method is its simplicity and ease of use. However, this approach tends to overestimate maintenance costs for large, expensive systems and underestimate for small, inexpensive techniques.

The second approach is to calculate maintenance labor and materials requirements individually according to not only the type of the control technique but also its size, complexity, ease of access, and other factors. As for operating labor, there are no general equations for calculating maintenance labor requirements. (However, Table 2.5 does provide factors for "add-ons.") Maintenance labor wage rates are available from the same sources that provide operating labor rates. If these data are unavailable, estimate the maintenance labor rate at 110% of the operating labor wage.[12]

Maintenance materials include small tools, lubricants, wire, tape, and other items consumed during the year. Their costs may be figured individually or, as a first estimate, as 100% of the maintenance labor cost.[13] *Do not* confuse maintenance materials with replacement parts. (See below.)

Replacement Parts

As stated above, "replacement parts" is a cost category not covered under the annual maintenance charge, for two reasons: (1) this cost is incurred less frequently than annually and (2) it is typically much larger in magnitude than the annual maintenance cost. Examples of control systems often requiring replacement parts are:

- catalytic incinerators: catalyst
- carbon adsorbers: carbon
- fabric filters: bags
- mist eliminators: pads
- thermal incinerators: refractory lining

Along with the cost of the replacement parts themselves (which should include applicable sales taxes and freight), the cost of the *replacement labor* must also be included. This is usually figured on an "as required" basis. For instance, to remove spent carbon from a carbon adsorber and replace it with fresh adsorbent may require a 2- to 3-person crew for 2 to 3 days. The cost of their time (including overhead), plus other costs (travel, lodging, meals, equipment rentals, etc.) comprises the replacement labor cost. As a first approximation, however, one can figure this labor cost at 100% of the replacement parts cost.[14]

Because the replacement parts cost is incurred less often than annually, it must be treated as a *capital* expenditure, with a "useful life" of its own. The following formula is used to calculate the "equivalent annual" replacement parts cost, C_{rp}:

$$C_{rp} = P_{rp}(CRF_{rp}) \tag{2.4}$$

where P_{rp} = cost of replacement parts and labor (including taxes and freight)

CRF_{rp} = "capital recovery factor" (defined in a later section of this chapter)

The CRF incorporates both the useful life of the replacement parts (usually 2 to 5 years) and the interest rate (or "opportunity cost") for the investment.

Finally, when figuring either the depreciation or the capital recovery for the control system, be sure first to deduct the replacement parts cost from the total depreciable investment. This is necessary to avoid double-counting.

Waste Treatment and Disposal

In some cases, the air emissions captured by the control system can be sold or recycled to the process, thus offsetting part or all of the total annual cost.

However, more often than not, the captured pollutants cannot be reused and therefore must be disposed of. Ultimately, disposal involves hauling the material to a landfill or other depository or burning it in a solid/liquid waste incinerator. For the former, hauling costs depend on the distance from the control device to the disposal site, the quantity of dust hauled, and so forth. Most hauling costs are figured on a $/ton-mile basis. In addition, a landfill charge ("tipping fee") must be paid for the privilege of dumping said material. However, federal regulations severely restrict the types of wastes that may be landfilled without "pretreatment" to remove hazardous substances.

Even pretreatment cannot make some wastes acceptable for land disposal; these must be incinerated. However, the cost of incineration can be quite expensive — $100/ton of waste or more — and is discouraged by the shortage of incineration capacity in some areas. (Chapter 5 will discuss liquid and solid waste treatment/disposal in more detail.)

Overhead

This indirect annual cost is arguably the easiest cost to compute, but among the most difficult to understand. The confusion surrounding it mainly stems from the several different ways it is computed and the many costs it includes, some of which appear to overlap.

Most estimators break overhead into two categories: *payroll* and *plant*. The former includes expenses associated with operating, supervisory, and maintenance labor, such as Social Security, Workmen's Compensation, and other taxes; health, life, and other insurance premiums; pension fund contributions; vacations; and other fringe benefits. Some of these costs are "fixed," in that they must be paid regardless of the number of hours an employee works in a given year. Nonetheless, payroll overhead is generally computed as a fraction of the total labor cost.

Plant overhead covers those expenses not necessarily tied to the operation and maintenance of a control technology, but which must be allocated in some way to it. These expenses include, for example, facility security, plant offices, parking area, lighting, cafeterias, locker rooms, etc. Like payroll overhead, plant (or "factory") overhead is figured by some as a fraction of the total labor cost (plus maintenance materials). However, other estimators compute it as a fraction of labor plus the total capital investment. The choice is an arbitrary one.

For most estimating purposes, it is both convenient and sufficiently accurate to combine both plant and payroll overhead into a single factor — a percentage of total labor (including maintenance materials). This is the approach recommended by Peters and Timmerhaus.[15] They recommend a range of 50 to 70% of total labor, a range that encompasses most figures in the literature.

Property Taxes, Insurance, and Administrative Charges

These indirect annual costs are "fixed" charges. Taxes and insurance are self-explanatory. "Administrative charges" encompass sales, research and development, accounting, and other home office (*not* plant) expenses.

Logically enough, taxes and insurance are typically figured from the total capital investment, as tax bills and insurance premiums are both based on the property valuation. By convention, "administrative charges" (also known as "G&A," "general works expense," etc.) are also factored from the TCI. Thus, it makes sense to lump them together into one figure. Although the value of this figure varies from firm to firm, one reference recommends 4% of the TCI, broken down as follows: property taxes — 1%, insurance — 1%, administrative charges — 2%.[16]

Depreciation and Capital Recovery

According to Katell and Humphreys, *capital recovery* is "the replacement of the original cost of an asset plus interest." But, they define *depreciation* as "a *form of capital recovery* applicable to a property with two or more years' lifespan, in which an appropriate portion of the asset's value is periodically charged to current operations" (emphasis added).[17] Thus, the notion of capital recovery is more general than depreciation, which is a form of capital recovery used for accounting or income tax purposes.

Grant et al. present a thorough explanation of the classic method used for calculating capital recovery.[18] In this method, we set aside a certain amount (or "payment") for each year of the project's life. Each payment will then earn interest (i) at the rate the firm deems "minimally attractive" for its investments. (For instance, in a five-year project, the payment made in year 2 will earn interest for three years.) The sum of these payments over the life of the project, plus the interest they earn, must equal the initial depreciable investment (less any salvage) plus the interest the investment would have earned had it been invested elsewhere at the same rate of return. This annual capital recovery payment (A) is calculated as follows:

$$A = (TDI - C_{rp})(CRF) + TNDI(i) \qquad (2.5)$$

where

$$
\begin{aligned}
TDI\text{-}C_{rp} &= \text{total depreciable investment } \textit{less} \text{ replacement parts} \\
&\quad \text{costs (see Eq. 2.4)} \\
TNDI &= \text{total nondepreciable investment} \\
&= \text{land + working capital + salvage} \\
CRF &= \text{capital recovery factor} \\
&= [i(1 + i)^m]/[(1 + i)^m - 1]
\end{aligned}
$$

Values of the CRF are tabulated in Grant et al. and other engineering economy texts. Note that "m" in the CRF formula denotes the *project* life, while "i" denotes the "minimum attractive" rate of return.

While engineering economists favor the capital recovery factor method, accountants prefer to use *depreciation* techniques for "recovering" the investment. More precisely, we should say "depreciation-plus-interest," because interest payments must be included here as well. As Grant et al. put it: "The total of depreciation plus interest is intended to serve the same purpose as [the] annual cost of capital recovery with a return."[19] Several depreciation methods are used: "straight line," "sum-of-years-digits," "declining balance," "sinking fund," and others. In fact, some firms use one method for internal accounting and another for calculating income tax deductions. Appendix A presents an overview of depreciation methods as they relate to income tax preparation.

Finally, keep in mind that regardless of whether the "capital recovery" or "depreciation" approach is used, the object is to obtain an annual expenditure that will correctly account for the initial investment plus an acceptable return.

MEASURES OF MERIT

When controlling an emission source, several control techniques may be available, each of them equally efficient. In such cases, cost would dictate the technique selected. There are several "yardsticks" of measures of merit that cost analysts use to compare the costs of competing control techniques. This section will cover these measures. Before doing so, we need to understand several cost engineering concepts and how they are used.

Time Value of Money

The old saying, "A bird in the hand is worth two in a bush," is the essence of this concept. According to Grant et al., "because of the existence of interest, a dollar now is worth more than the prospect of a dollar next year or at some later date." Therefore, *interest* (or "opportunity cost") is the measure of the time value of money, or ". . .the return obtainable by the productive investment of capital."[20]

In selecting the interest rate to use in a cost analysis, one must distinguish between *nominal* and *real* rates. *Nominal* interest rates are the usual bases for computing periodic interest payments. The interest rate charged by a lending institution on a loan is a nominal rate, as are most of the rates reported by the financial media (e.g., the "prime lending rate"). Nominal rates are used in cost analyses when the magnitudes of the costs are subject to inflationary pressures. Such analyses are said to be done in *nominal* dollars. Example: in year 2, the labor cost for a control system may be $10,000, but due to wage hikes tied to the Consumer Price Index, the labor cost is 5% higher in year 3, or $10,500. In such an analysis, the nominal interest rate might be 10%.

Conversely, *real* or *constant-dollar* cost analyses do *not* consider inflation. In constant-dollar analyses, the only year-to-year cost changes one would see would be due to *real* increases in costs due to *market* (not inflationary) forces, such as the high gasoline prices brought about by the 1970s oil embargoes. Accordingly, only *real* interest rates should be used with these analyses.

The next equation relates real and nominal interest rates:

$$(1 + i_n) = (1 + i)(1 + r) \qquad (2.6)$$

where   i, i_n = real and nominal interest rates, respectively
        r = annual inflation rate

Wherever possible, constant-dollar cost analyses should be made, because of their simplicity. "Real" analyses are especially well-suited to analyses where income taxes are not considered—i.e., those done for governmental-type projects, such as public works, regulatory impacts, and the like. But where income taxes are considered, "nominal" analyses might be more feasible, if for no other reason than because depreciation deductions are made in nominal dollars, while taxes are figured on the same basis.

If the cost analysis noted above were done in constant dollars, the year 3 labor cost would still be $10,000 (i.e., no inflationary increase from year 2 to 3). Further, based on equation 2.6, the real interest rate would be 4.8%, given a 5% inflation rate.

Cash Flow

During the lifetime of a control technology, expenditures are made and (if applicable) revenues are received. The capital cost-related expenditures typically occur at (or before) the startup of the system, while in every year thereafter, the various annual costs are spent. The amounts and timing of these costs and revenues comprise the project *cash flows*. How does one calculate the *net cash flow* (NCF) for a given year? First, if income taxes are *not* considered *and* the analysis is done in "real" dollars, the NCF is simply:

$$NCF = Revenues - Costs = TAC^* \qquad (2.7)$$

Here, TAC* is simply the sum of the direct and indirect annual costs described above, *except for* the capital recovery charge. If costs exceed revenues in a year, the NCF will be negative for that year—usually the case for control systems.

Prototypical "cash flow diagrams" are shown in Figure 2.1. Each diagram shows the NCF for every year from the beginning to the end of a hypothetical five-year project. Note that each NCF occurs at the *end* of the year in question, even though, in reality, payments will be made unevenly during the year. This convention, recommended by Grant et al. and others, simplifies calculations. Moreover, the error introduced by this assumption is minimal compared to the precision of the cost estimates themselves.

The first diagram in Figure 2.1 shows six net cash flows: the TCI, made in

Diagram I. Constant (real) dollar analysis, <u>no</u> income taxes . . .

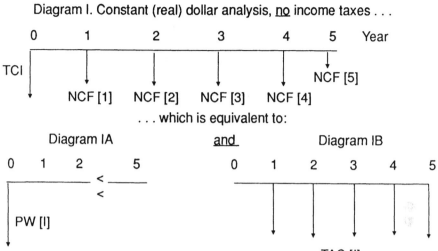

. . . which is equivalent to:

Diagram IA   <u>and</u>   Diagram IB

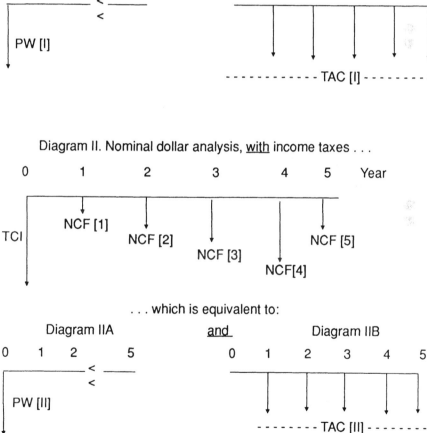

Diagram II. Nominal dollar analysis, <u>with</u> income taxes . . .

. . . which is equivalent to:

Diagram IIA   <u>and</u>   Diagram IIB

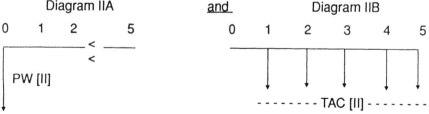

Figure 2.1   Example cash flow diagram.

year "0," and one NCF (TAC*) for each of the five years of the project. Notice that the TAC*s are exactly equal from year 1 to year 4. This reflects the fact that this is a "constant-dollar" analysis, so that the costs and revenues are unaffected by inflation. In year 5, however, the NCF increases (i.e., becomes less negative). Why? At the end of year 5, the end of the project, the working capital, land, and equipment salvage are added back in as "revenues," thus decreasing the (negative) magnitude of NCF_5. This first diagram (I) is somewhat idealized, because it doesn't factor in inflation or income taxes. How would taxes change matters? First, we would need to expand our definition of NCF to incorporate depreciation (Dep) and the combined federal-state-local income tax rate (t):

$$NCF = (1 - t)(TAC^*) + t(Dep) \qquad (2.8)$$

This is the general form of the NCF equation. Clearly, if income taxes were not considered, $t = 0$, and Equation 2.8 would reduce to Equation 2.7.

Notice that depreciation appears in Equation 2.8 as a revenue, rather than a cost. Why? Because here depreciation has been used for income tax purposes only, not for calculating the capital recovery charge. Further, the income tax is figured against the *net taxable income*, which equals revenues minus costs *minus* depreciation (i.e., TAC* – Dep). And since in this case depreciation is not a true out-of-pocket expense, it is *added back in* after being subtracted out to obtain the net taxable income. (For a fuller discussion of cash flow, see Uhl.[21])

The second cash flow diagram (II) in Figure 2.1 depicts a project in which both inflation and income taxes have been considered in the analysis. Notice that the NCFs decrease gradually (become more *negative*) from years 1 to 3, thanks to inflation. The NCF plummets in year 4 but then increases sharply in year 5. This apparently strange behavior is caused by two things: first, in accordance with tax rules, the investment is totally depreciated by the end of *year 3*. Thus, in year 4, the "t(Dep)" term is 0, and the NCF drops. Finally, as in diagram I, the nondepreciable investment items partly offset the "(1 – t)TAC*" in year 5, resulting in a higher (less negative) NCF.

Present Worth Measure

After calculating the NCF for each year, how can we put them all "on an equal footing," so that we can compare the NCFs of one project to the NCFs of another? One way is to calculate the *present worth* of the projects. This entails "discounting" all cash flows occurring after year 0 (the year in which the technology is placed in service) back to year 0. These NCFs are discounted by multiplying each by a "discount factor" $(1/[1 + i]^k$ or $1/[1 + i_n]^k)$ where "k" is the number of years from year 0 to the end of year k, the year in which the cash flow is incurred. Here also, i or i_n are the appropriate real or nominal interest ("discount") rates, respectively. The sum of these net discounted cash flows

(NDCF), when added to the total capital investment, yields the "net present worth" of the project. When comparing control technologies, the one with the highest positive (or lowest negative) present worth is the one to select. However, when using the present worth method, *never* compare technologies with different useful lives (e.g., 10 vs. 15 years) or different economic bases (e.g., real vs. nominal-dollar analysis). If this is done, the comparisons will be invalid.

When making present worth comparisons where income taxes are included, one must also take care that the interest rate used in the discounting is an *after-tax* rate. For example, if a firm's before-tax (nominal) return is 10% and its combined marginal income tax rate is 50%, its after-tax return would be 5%. To do otherwise would be to bias the comparison results.

Cash flow diagrams IA and IIA in Figure 2.1 show the results of two present worth calculations. Each set of NCFs has been "shrunk" to a single present worth (PW) via the discounting procedure described above. Note that to obtain PW(I) a *real* interest rate was used, while a *nominal, after-tax* rate was used to calculate PW(II).

Internal Rate of Return Measure

As stated previously, the interest rate used in the discounting calculation is typically the "minimum acceptable" before- or after-tax rate of return used by firms in making their investment decisions. In some comparisons, cost analysts do *not* select an interest rate for the discounting calculations. Instead, they arbitrarily set the present worth of each alternative equal to 0 and then solve (usually via trial-and-error) for the interest rate that would solve the equation. This solution is the *internal rate of return* (IRR). Then, the alternative having the highest positive (or least negative) IRR is the one selected. Thus, the IRR is a special case of PW analysis. And like PW, the IRR requires that all alternatives have equal lives and equivalent economic bases.

Equivalent Uniform Annual Cash Flow (EUAC) Measure

Of all the "measures of merit," the EUAC method is arguably the most commonly used, by virtue of its versatility and ease of use. Moreover, the EUAC may be used to compare alternatives having different lives. It is not very efficient when treating uneven cash flows over the life of a project, such as in diagram II of Figure 2.1. However, in those cases where cash flows are constant, the EUAC method is recommended. If in the diagram I illustration, the NCF in year 5 had equalled the NCF in years 1 to 4, the total annual cost (TAC_I) would have been:

$$TAC_I = TAC^*_I + A_I \qquad (2.9)$$

where A_I = capital recovery payment (see Eq. 2.5)

The "A_i" term in this equation "smoothes out" the total capital investment over the project life. When added to the TAC* terms (which are equal in *constant* dollars), the TAC results.

The EUAC method cannot be directly applied to the diagram II cash flows. However, EUAC *can* be used to convert the present worth, PW_{II}, into equal annual payments, as follows:

$$TAC_{II} = PW_{II} \times CRF_{II} \tag{2.10}$$

Of course, CRF_{II} must incorporate a *nominal* interest rate, because the diagram II illustration is in nominal dollars. Finally, diagram IIB in Figure 2.1 shows the results of this smoothing process. But again, we cannot compare TAC_{II} with TAC_I, as they reflect different costing assumptions.

Other Measures of Merit

Along with the three methods just discussed, others have been used in economic comparisons. These are: (1) return on investment (ROI), (2) payout time, and (3) unit costs. Because none of them is as flexible and accurate as PW, IRR, and EUAC, they are not recommended. Nonetheless, because of their widespread use, they deserve passing mention.

Return on Investment

This measure is just the quotient of the net profit and the total capital investment (including land and working capital). The net profit is the net cash flow *minus* depreciation. But because ROI does not incorporate a rate of return, it ignores the "time value of money," the very cornerstone of the engineering economy discipline. When applied to long-term projects, ROI can introduce significant errors. But, for projects with short lives (less than five years), ROI can give "quick-and-dirty" results.

Payout Time

Analogous to ROI, payout time is the period (in years) in which the accumulated net cash flows equal the *depreciable* investment. (Some analysts mistakenly assume that payout is just the reciprocal of ROI, but that is incorrect.) Like ROI, the payout time calculations suffer from the exclusion of the requisite interest rate.

Unit Costs

According to Uhl,[22] "Unit costs are costs per unit of product, service, or output." He lists four kinds: (1) operating expense per unit of product (e.g., mills/kwhr), (2) capital investment per unit rate of output or capacity (e.g., $/kw), (3) unit cost for cost-effectiveness analysis, and (4) revenue requirement per unit of output.

The first three are easy to understand and compute, though like ROI and payout time, they do not incorporate interest rates. The "revenue requirement" (RR) method can or cannot involve discounting, but either way, it can be complex to calculate. If discounting is used, the revenue requirement per unit output is termed "levelized cost." RR is most often used with public utility cost analyses. (For more information on "unit costs," see Uhl.)

TYPES OF COST ESTIMATES

In the preceding discussion, we've made no mention of the *accuracy* of the costs. Alas, it's difficult to determine the accuracy of an estimate, owing to the several different kinds of costs that comprise a capital or annual cost estimate, the different assumptions that go into calculating *each* of these costs and so forth. Nevertheless, for capital costs at least, accuracies have been assigned to estimates, based on the types and amounts of data that went into making them. The American Association of Cost Engineers lists three estimate classes:[23]

- *Order-of-Magnitude* (accuracy: +50% to -30%): This is ". . .an approximate estimate made without detailed engineering data." It is the least accurate of all estimates, yet the simplest to make. An example would be where the TCI of a control system is estimated to vary proportionately with the system capacity raised to some power (e.g., 0.6—the famous "six-tenths" rule).

- *Budget* (accuracy: +30% to -15%): This estimate is prepared ". . .with the use of flowsheets, layouts, and equipment details." Another reference breaks this class of estimates into two subcategories: "study" (±30%) and "scope/preliminary" (±20%).[24] To make a control system "study" estimate, for example, one would need to know the location of the emission source and have a rough sketch of the system layout. In addition, the sizes and types of system equipment items, ductwork, and buildings would have to be known, along with utility requirements. (Chapter 3 covers these system design topics in detail.)

  "Scope/preliminary" estimates require the same kinds of data inputs as study estimates, only in more detail. Also required are insulation and instrumentation specifications.

- *Definitive* (+15% to -5%): The *definitive* are the most accurate class of estimates, but the most expensive to make. The minimum inputs include plot plans, piping and instrument diagrams, a complete set of specifications, and other engineering data. At its most "definitive," this type of estimate would be acceptable for purposes of soliciting and evaluating contractors' bids. The cost data and procedures in this book are mainly geared to making "budget-study" estimates, as they represent a compromise between the crude "order-of-magnitude" estimates and the accurate (but expensive and very detailed) "definitive" estimates. Moreover, for comparing the costs of control technologies, formulating emission control strategies, and other purposes for which this book is intended, the ±30% accuracy of the study estimate is quite sufficient.

*   *   *

In this chapter, we've described the various kinds of costs that make up a cost estimate, presented methods for analyzing and comparing these costs, and discussed their accuracy. In following chapters, we'll apply this knowledge to the sizing and costing of specific control technologies. But first, we need to get a better understanding of these technologies—what they are, how they're "put together," and how they're used. That's the purpose of the next chapter.

REFERENCES

1. Humphreys, K. K., and S. Katell. *Basic Cost Engineering*. New York: Marcel Dekker, 1981 (hereinafter cited as *Basic Cost*).
2. Malstrom, E. M., Ed. *Manufacturing Cost Engineering Handbook*. New York: Marcel Dekker, 1984.
3. Humphreys, K. K., Ed. *Project and Cost Engineers' Handbook*, 2nd ed. New York: Marcel Dekker, 1984.
4. Grant, E. L., W. G. Ireson, and R. S. Leavenworth. *Principles of Engineering Economy*, 7th ed. New York: John Wiley and Sons, 1982 (hereinafter cited as *Engineering Economy*).
5. Vatavuk, W. M., and R. B. Neveril. "Estimating Costs of Air-Pollution Control Systems, Part II: Factors for Estimating Capital and Operating Costs," *Chemical Engineering*, November 3, 1980, pp. 157–162 (hereinafter cited as "Costs, Part II").
6. Uhl, V. W. *A Standard Procedure for Cost Analysis of Pollution Control Operations* (Volumes I and II). Research Triangle Park, NC: U.S. Environmental Protection Agency, June 1979 (EPA-600/8-79-018b) (hereinafter cited as *Standard Procedure*), pp. A15-A16.
7. *Basic Cost*, pp. 31–32.
8. *Basic Cost*, pp. 31–32.
9. "Costs, Part II."
10. Peters, M. S., and K. D. Timmerhaus. *Plant Design and Economics for Chemical Engineers*, 3rd ed. New York: McGraw-Hill, 1980 (hereinafter cited as *Plant Design*), p. 195.
11. "Costs, Part II."
12. "Costs, Part II."
13. "Costs, Part II."
14. "Costs, Part II."
15. *Plant Design*, p. 203.

16. Calvert, S., and H. M. Englund, Eds. *Handbook of Air Pollution Technology*. New York: John Wiley and Sons, 1984, p. 341.
17. *Basic Cost*, pp. 101, 105.
18. *Engineering Economy*, pp. 34–37.
19. *Engineering Economy*, p. 211.
20. *Engineering Economy*, p. 23.
21. *Standard Procedure*, pp. C1-C7.
22. *Standard Procedure*, pp. E5-E7.
23. *Basic Cost*, p. 108.
24. Perry, R. H., and C. H. Chilton. *Perry's Chemical Engineer's Handbook*, 5th ed. New York: McGraw-Hill, 1973, pp. 25–12 to 25–16.

CHAPTER 3

At the Drawing Board: An Estimating Methodology

Though this be madness, yet there is method in 't.

Hamlet — William Shakespeare

At first glance, estimating the cost of an air pollution control method would seem to be an easy task. After all, what's so hard about selecting the equipment for controlling a source and then estimating what it would cost to buy and run that equipment? You don't need to be an expert in, say, fabric filter design and operation to know that you would need one to control a particulate emissions source. Nor does it take any expertise to compile a few facts about that source and to give this information to a baghouse vendor (manufacturer) who would use it to develop a cost quote.

True enough. But one must know at least *something* about emission sources, equipment selection and sizing, and cost estimating to be able to give a vendor the *right* kinds of information. Moreover, vendors may not be able to tackle the entire job. Some functions, such as sizing certain kinds of auxiliary equipment (e.g., ductwork), the customer will have to perform. And even if a vendor can "do it all," the customer should know as much as possible about the product he/she is buying, especially if the customer is a technical person who will be responsible for installing, starting up, and operating the control system.

That's what the rest of this book tries to do — enable prospective buyers and operators of air pollution control systems to make more intelligent decisions in that area. In Chapter 2, we explored the mysterious land of cost engineering, learning about what kinds of expenditures are relevant to air pollution control technologies, how those expenditures are assembled to produce capital and annual cost estimates, and how to accurately and consistently compare the costs of competing control technologies. In this chapter we'll develop a general methodology for selecting, sizing, and costing controls. In following chapters, we'll apply this methodology and the material covered in Chapter 2 to specific kinds of control methods. But first, let's discuss these methods.

COLLECT, CONVERT, CONTAIN, OR PREVENT?

Broadly speaking, there are four ways to control real (or potential) air emissions from stationary sources: (1) *collect* them, (2) *convert* them, (3) *contain* them, or (4) *prevent* them from having the opportunity to be emitted in the first place. These methods can be applied to both "outdoor" and "indoor" emission sources.

"Add-on" control devices are the typical means for collection and conversion. These are what most envision when thinking about air pollution control. "Add-ons" are usually installed at the tail end of a source after the emission "capture" device. A capture device can be a hood, a total enclosure (such as a furnace building), or simply a duct connecting the source to the collection/conversion device. "Collection" devices include fabric filters, electrostatic precipitators (ESPs), wet scrubbers, carbon adsorbers, and refrigerated condensers. "Conversion" devices change the *form* of the emissions in some way, typically via combustion to carbon dioxide, water, hydrogen chloride, and other compounds. Combustion devices include catalytic incinerators (fixed- and fluid-bed), thermal incinerators (recuperative and regenerative), and carbon monoxide boilers.

Combustion is not the only means used for converting emissions, however. For example, in a flue gas desulfurization system (FGD), sulfur dioxide reacts with limestone or other minerals in a gas absorber to form calcium sulfate or other solid products. Common to all "add-ons," however, is the fact that their installation and operation rarely *affects* the quantity or composition of the emissions generated by the source.

This cannot be said of the other two categories—"containment" and "prevention" methods. "Containment" methods do exactly that: contain the pollutants within the source, to prevent their release to the outer (or inner) atmosphere. What are some containment methods? A vapor balance system used to contain gasoline emissions during tank truck loading/unloading operations is one. So is an internal floating roof installed inside a fixed-roof crude oil storage tank to inhibit volatilization and release of organic compounds. Other containment methods are particular to the sources they control, such as membrane covers installed on surface impoundments at hazardous waste treatment facilities. (The latter containment method is often used in conjunction with a collection/combustion device to ultimately dispose of the emissions.)

But whether the emissions are contained, collected, or converted, they must be *dealt with* in some way. Collected emissions, for example, often require additional treatment and handling before they can be disposed of or, if they have value, recycled or sold. And this treatment and handling adds to the cost of capturing and collecting the original emissions.

Furthermore, disposal of collected emissions can create additional pollution problems in *other* media. Consider a mixture of organic solvents captured by a carbon adsorber. Before the advent of the Resource Conservation and Recovery Act (RCRA) and other hazardous waste control legislation, these solvents might have been collected in drums and then simply dropped at a landfill. But not any more. If the solvents belong to a "restricted" category, they cannot be disposed of on land without prior treatment. To date, the U.S. Environmental Protection Agency, which administers the RCRA, has restricted the land disposal of dioxins and certain solvents and has banned outright the dumping of biphenyls, cyanides, and metals. Moreover, as this is being written, EPA is proposing to widen these restrictions to 24 types of hazardous wastes from petroleum refineries, steel mill emission-control systems, and other plants.[1] Thus can pollution control be an ever-widening (and seemingly vicious) circle.

Of the four control categories, "prevention" methods can be the most attractive, for they seemingly "solve" an emission problem before it is created. Prevention methods cover a wide range of techniques, most of which involve *process modifications* of some kind. These modifications, in turn, may involve changes to the *process equipment* and/or changes to the *operating materials.* Consider a coating operation that consists of conventional spray booths in which organic solvent–borne paints are applied. If this spray equipment were replaced by a "powder" coating process (in which the coating is applied to the parts without the use of any solvent), the prevention method would be a process equipment change. However, if a waterborne or other low-solvent coating were substituted for the high-solvent coating *without* any changes to the process equipment, it would be an operating material change only.

Another class of operating materials changes is *fuel substitutions.* Typically, this entails burning low-sulfur fuel in lieu of fuel with a higher sulfur content, thereby reducing the quantity of sulfur dioxide emitted. The fuel substitution technique is often used to reduce SO_2 emissions from electric utility, industrial, and other boilers.

Common to all of the prevention techniques is the *a priori removal of potential air emissions from the source.* That is, the higher-sulfur fuel never gets burned, the higher-solvent coating never gets applied, and, consequently, the emissions never get generated.

Finally, there is another, rarely-thought-of emission prevention method: *production curtailment.* This can mean anything from a partial cutback to a total shutdown. By reducing the "activity level" of the source (i.e., the amount of material processed or the hours of operation), the emissions generated per unit time can also be reduced. The greater these activity reductions, the lower the total emissions. Needless to say, the minimum emissions would be realized when the process were no longer in operation. However, because the plant shutdown "control alternative" can have many undesirable non-environmental effects (e.g., job losses), it is only used as a last resort. But in the event of a total shutdown, a plant may still emit some pollutants — such as "fugitive" particulate from abandoned storage piles — even though the formerly emitting processes are no longer operating.

STEP 1: GETTING TO KNOW THE EMISSION SOURCE

Knowing the different methods we can use to control a source is just the beginning, however. We next need to collect all the information we can about the source to be controlled. The more data we can find at this point, the better we can define the project for purposes of soliciting vendor cost quotes. What *kinds* of data? In short, *any* information that would relate to the sizing and costing of the control system. For an "add-on" controlling a process vent, for instance, the source data would include:

- waste gas stream characteristics — volumetric flowrate, temperature, pressure, and composition (in particular, moisture content)
- emission characteristics — mass rate(s), particle size distributions, etc.
- operating schedule (e.g., hours/year)
- status (new or existing)
- controls already in place
- geographic location

But if the emission source were, say, a solids storage pile to be controlled by a wet dust suppression system, the kinds of data collected would be different. They would include such information as the dimensions of the pile, its moisture content and particle size distribution, the prevailing meteorological conditions, and like information. Because the kinds of source information to be collected vary a great deal, there is no "laundry list" that would apply in all cases.

Before leaving this topic, we need to say a few words about the last three "bullets" in the above list. First, why should the sizing and costing of an "add-on" control system be affected by the "status" of a source? After all, isn't the control device sized to control a given waste gas stream, and wouldn't the characteristics of that stream be the same in either a new or an existing plant? Indeed, they would — but the source status likely *would* affect the type and size of auxiliary equipment used with the control device, as well as the system installation costs. In an existing plant (or "retrofit" installation), for example, more ductwork might be needed to connect the source with the control device than in a new ("grass roots") plant. Similarly, if a control device, such as an electrostatic precipitator, were already in place to control the source, one might opt to upgrade the ESP instead of replacing it with an entirely new ESP or another device (say, a fabric filter). The cost savings here would be more than trivial. Finally, the location of a source would affect not only the system design (e.g., extra insulation for extreme climates), but the cost of materials, utilities, and other commodities.

In following sections, we will discuss retrofit costs and related subjects at greater length. For now, keep in mind that a control method's design and cost rest on the foundation of source data. The more complete and reliable these data, the more solid the foundation.

STEP 2: SELECTING AND CONFIGURING THE CONTROL EQUIPMENT

"There is more than one way to skin a cat," goes the old saw. The saying applies to air pollution control system design as well. For there may be several equally effective ways to control a given emissions source. Choosing the best "control option" for a source usually involves sizing and costing the alternative methods and selecting from these the one with the lowest annual cost. But before the sizing and costing comes the *selecting*. And this simply entails choosing the equipment that comprise the control system, be it a traditional "add-on" or a nontraditional process change. However, because process equipment and operating materials changes are too varied to generalize, we'll confine the discussion in this section to "add-ons" only.

First of all, along with the control device itself (and built-in instrumentation), every add-on system contains certain kinds of common equipment:

- *Capture device* — a hood, duct, or other means for collecting the emissions at the source
- *Ductwork* — for conveying the emission stream from the capture device to the control device
- *Fan system* — for moving the emission stream through the control system
- *Stack* — for dispersing the cleaned gas to the atmosphere

In addition, add-on systems that *collect* emissions also usually require some means for treating/disposing of the captured pollutants. For dry collection systems, this would likely mean a dust hopper; a screw, pneumatic, or other type of conveying system; and means for transporting the dust to a disposal site or back to the process (if it is recycled). With wet systems, it becomes more complex. Here, the captured emissions are in a water stream and must be treated for removal or conversion of the pollutants. For instance, with some venturi scrubber systems, the wastewater is clarified and vacuum filtered to remove the solids. The resulting filter cake is then ultimately disposed of at a landfill (*if* it is deemed nonhazardous by RCRA).

Some add-on systems may also require pretreatment devices to reduce the pollutant loading or decrease the stream temperature or moisture content before the offgas reaches the control device. For instance, relatively inexpensive mechanical collectors (cyclones) are often used upstream of fabric filters, ESPs, and other "dry" collection devices to remove most of the larger (and many of the smaller) particles from the waste stream. Similarly, spray chambers or quenchers are occasionally used ahead of wet collectors (such as venturi scrubbers) to cool the offgas by adding evaporative moisture to it. The overall effect of this is to reduce the stream volumetric flowrate and, hence, the size and cost of the control device. Clearly, evaporative coolers should not be used ahead of dry collectors, because the added moisture can cause the particles to agglomerate, clogging the bags in a baghouse or gumming up the collecting plates in a dry ESP.

How this equipment is arranged, or *configured*, in the control system has a bearing on the size and cost of the individual pieces. For example, the system fan can be placed on the upstream (pressure) or downstream (suction) side of the control device. If placed upstream, the fan might have to convey a hotter — and hence, higher-volume — waste gas stream. If so, a larger (and more expensive) fan would be needed. Moreover, in pressure configurations a fan would be handling a stream with a higher pollutant concentration. This could dictate the use of special erosion or corrosion-resistant materials of construction. In the case of a stream with a high particulate concentration, a "radial-tip" fan may be needed. Radial-tip fans are more tolerant of high particulate loadings than "backward-curved" and other commonly used centrifugal fans. Does this mean that the fan should *always* be placed on the suction (downstream) side of the control device? Not at all. There may be situations where the control device — say, a baghouse — would have to be reinforced to withstand negative operating pressures, even just a few inches of water.

The system configuration will also affect the size and cost of the ductwork. Consider a venturi scrubber that is preceded by a spray chamber for cooling the incoming waste gas. Further assume that the venturi has to be located 150 feet from the emission source. If the spray chamber were installed close to the source, the temperature (and hence the volume) of the waste gas stream would be reduced far upstream of the venturi. Thus, a smaller-diameter (and less costly) duct would be needed. Conversely, there may be cases where it would be more economical (e.g., piping costs) to install the spray chamber closer to the venturi. These are just two of several factors that should be considered when configuring the control system. The final configuration would resemble the flowsheet shown in Figure 3.1. The duct lengths and diameters are given, as are the waste gas temperatures and flowrates. (The duct diameters have been calculated via the procedure shown in Chapter 4.) Note how the duct diameter decreases from 30 to 26 inches between the source and the stack, while the gas temperature drops from 500 to 150°F. Clearly, had the spray chamber been placed closer to the venturi than to the source, the gas temperature (and duct diameter and cost) would have been much higher in Section A. This is not a trivial consideration, as the ductwork cost can comprise a significant part of the purchased equipment cost.

Figure 3.1 illustrates several other facets of control system configurations. First, most systems are designed to operate at (or very near) atmospheric pressure. Hence, the Ideal Gas Law may be used to calculate gas volumes. And notice that the gas flowrate (in scfm — standard cubic feet per minute, taken at 70°F and 1 atmosphere) increases from 11,000 at the source to 12,700 at the stack. This increase is due to the water added in the spray chamber and venturi. By the time the gas leaves the venturi, it is saturated. (According to the psychometric chart, the 500°F inlet air with a 15 volume percent moisture content would reach saturation at 150°F, if it were cooled adiabatically.) Next,

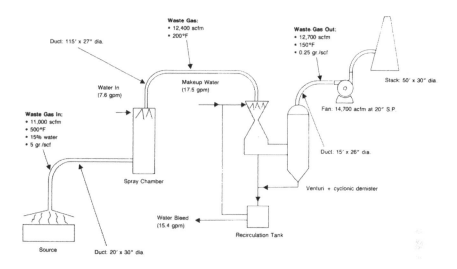

Figure 3.1 Hypothetical control system configuration.

note that the venturi outlet temperature (150°F) is too low to provide much "lift" to the flue (stack). In fact, a 300°F minimum stack temperature is recommended. In such a case, a "reheater" might be used to give the flue gases the required thermal boost.

Finally, note that the fan has been placed *after* the control device (the venturi). This serves to reduce the volume conveyed by the fan. Also, because the fan is handling a relatively clean gas (0.25 grains/scf = 5% of the inlet loading), erosion of the blades and other moving parts will not be a concern. The only drawback is that the entire system will be under up to 20 inches (water) suction. Although this may not affect the spray chamber or venturi, it *could* cause problems with the ductwork, which could buckle under excessive negative pressure. To avoid the suction problem, we could install a radial-tip fan ahead of the spray chamber, but as Figure 3.1 shows, it would have to convey a much larger gas flow—and would consume that much more electricity.

STEP 3: EQUIPMENT SIZING

Once the control device and auxiliaries are selected and configured, they must be *sized*. For each item, sizing entails choosing or calculating a *primary sizing parameter(s)*. With the waste gas volumetric flowrate, this parameter primarily determines the equipment size and cost. Table 3.1 lists primary sizing parameters for selected devices and auxiliaries. For some equipment, the primary sizing parameters define the gross dimensions of the item, while for others the parameter determines the size of some important *part* of the equipment (e.g., plate area). In any event, Table 3.1 lists only the *primary sizing*

Table 3.1 Primary Sizing Parameters for Selected Control Devices and Auxiliary Equipment

Equipment	Primary Sizing Parameter(s)
Control Devices	
Fabric filter	Bag area
Electrostatic precipitator	Collecting plate area
Wet scrubber (e.g., venturi)	Gas velocity, pressure drop (OR gas volume flow)
Carbon adsorber	Carbon weight, gas velocity (OR gas volume flow)
Gas absorber	Column height, diameter
Thermal and catalytic incinerators	Combustion chamber volume, heat exchanger area (OR gas volume flow and percent heat recovery)
Refrigerated condenser	Cooling load (OR gas volume flow, inlet/outlet temperatures)
Flare	Flare height, tip diameter (OR gas *mass* flow)
Auxiliaries	
Ductwork	Length, diameter (OR weight)
Fan	Wheel diameter, static pressure
Pump	Head, liquid flow
Stack	Height, diameter
Screw conveyor	Length, diameter
Cyclone	Gas volume flow (OR inlet area)
Spray chamber, quencher	Gas volume flow (OR length, diameter)
Plant buildings	Floor area

parameters. To cost a piece of equipment one may also have to determine the type and dimensions (e.g., thickness) of the construction material and insulation (if any), extent of instrumentation (including monitoring devices), and similar specifications. The primary sizing parameters are by far the most important in characterizing equipment.

What are the *magnitudes* of these parameters and how are they used? First, it would take dozens of pages to compile a comprehensive list of parameters for all equipment for all applications. Abbreviated listings are provided in Chapters 4 through 6, however, for the equipment covered therein. Some of these parameters are expressed on "normalized" bases. That is, instead of giving, say, the total bag area for a baghouse, the parameter shown is the *air-to-cloth* ratio — a ratio of the waste gas volumetric flowrate to the area of the bags. Also, a range of parameters may be listed, if they should vary according to the control device design, as well as the application.

Some of the primary sizing parameters have been based on actual experience with control applications. The air-to-cloth ratios, wet scrubber pressure drops, and carbon adsorber working capacities are examples. Other parameters have been computed from mathematical models based, in turn, on a mix of theory and experience. Regardless of how they were determined, the sizing parameters are sufficiently accurate for making "budget-study" type estimates.

But one cannot size a control device with these parameters alone. Another input is needed, something to "bridge the gap" between the sizing parameter and the equipment size. The waste gas volumetric flowrate provides that "bridge." For instance, if we wanted to size a fabric filter, we would divide the waste gas flowrate (in actual cubic feet/minute [acfm]) by the air-to-cloth ratio (in feet/minute [fpm]). This yields the *bag area* (in square feet), the variable against which the baghouse equipment cost is typically correlated.

> **Example**: To control particulate emissions from a foundry shakeout operation we would need a shaker-type baghouse with a 5.0-fpm air-to-cloth ratio. The waste gas flowrate is measured at 50,000 acfm. The bag area of the baghouse would therefore be: 50,000 acfm/5.0 fpm = 10,000 sq ft.

Lastly, some sizing parameters have a greater bearing on *annual* costs than on the equipment costs. For example, the wet scrubber design pressure drop (inches of water, typically) does affect the unit's operating pressure and, hence, its equipment cost. But pressure has an even greater influence on the *electric power* consumption and expenditure. In some cases, the power cost contributes well over half the control system annual cost.

STEP 4: ESTIMATING THE CONTROL SYSTEM COSTS

Chapter 2 described the components of the total capital investment (TCI) and total annual cost (TAC). Some of these components apply to all control systems; others to just a few. In most cases, the estimator will be able to fathom which is which, but in others, the choices may not be so obvious. In any event, the sections covering the individual control devices and auxiliaries will make the decisions easier. (See Chapters 4 to 6.)

SOURCES OF COST INFORMATION

The *purchased equipment cost* is common to all control devices. As Chapter 2 showed, this cost is comprised of the cost of the control device, plus the costs of auxiliary equipment, taxes, freight, and instrumentation. In turn, the costs of the control device and auxiliaries are the foundation of the purchased equipment cost. As alluded to above, these equipment costs correlate directly with the equipment sizes. In Step 3, we showed how to calculate equipment sizes from the sizing parameters, waste gas flowrate, and other inputs. But where can we find the equipment *prices*—or better yet, the correlations between them and the equipment sizes? There are three places to look:

1. Air pollution control equipment vendors
2. Owners and operators of control systems
3. Technical publications

The Vendors

Vendors are probably the best sources of price data—simply because in nearly all cases they are the ones who would supply the control devices to firms installing them. There are, literally, hundreds of vendors. (For a quite comprehensive listing, see *Pollution Equipment News 1988 Buyer's Guide.*[2]) These vendors manufacture all sorts of control devices and auxiliaries, from (gas) absorbers to (fan) variable speed drives. Because there are so many vendors, the air pollution control equipment business is intensely competitive. And due to this competition, vendors are usually reluctant to divulge their price data to just anyone asking for them. Vendors will provide cost quotations to inquirers, if they are given adequate specifications for the proposed control system. What do these specifications consist of? The "facility parameters" listed in Step 1 above provide a good start. In addition, before giving a quote some vendors may ask for such information as the pressure and quality of steam available onsite, the amount and temperature of cooling water available, the space available to erect the control equipment, et cetera. But in most cases, the facility parameters are sufficient for obtaining quotes.

In contacting vendors, two approaches are possible. Those needing cost data quickly can telex or telephone vendors. Although this is the fastest way, it has its drawbacks. For one, many vendors simply will not provide quotations over the phone. *Period.* Others may give costs orally, but these quotes are typically of the "order-of-magnitude" class — i.e., costs in terms of $/cfm or the equivalent. However, estimators usually need more accurate costs to develop "budget-study" estimates — quotes at least accurate to within ±30%.

Thus, if you have the time (if not, *make* the time!), send vendors written requests for quotations. Not only does a letter clearly detail the specifications for the control device, it lets the vendor know that yours is a serious request. He/she will also know at once whether you are requesting an "informational" or a "job" quote. Anyone not planning to buy a control device would fall in the former category. If you are requesting a job quote, however, the vendor will get to it promptly and will usually provide a full quotation within one to three weeks, depending on how busy he or she is.

An informational request may take much longer to process. Why? Simply because the vendor stands to gain absolutely nothing from supplying you cost information, save for a measure of good will — hardly a bankable commodity these days. *If* a vendor should reply to an informational request, he or she likely will squeeze it in between preparing bids for (potential) customers. This does not reflect hardheadedness on the vendor's part, but merely good business practice.

Whether informational or job-related, a vendor quotation will usually indicate a period of validity (typically 30, 60, or 90 days). This is standard policy for many vendors, meant to protect them against the vagaries of the marketplace. In reality though, the prices may not change for some time, so that there likely is no need to obtain an entirely new set of quotes every two or three months. Also keep in mind that most vendor quotations are "F.O.B." Again, this is standard operating procedure. "F.O.B." ("free-on-board") just means that the cost of freight must be added to the prices shown.

If the estimator is fortunate, a vendor will send a price list, giving prices for *all* of the models he/she manufactures, not simply for the ones requested.

UNIVERSAL AIR PRECIPITATOR

"Mighty Ion"
Self-Contained Industrial Products

| Effective: Oct. 15, 1987 | Base Price Schedule | | Price List MI-7 |

Models	Capacity (cfm)	Motor	Base Price
MI 8	800	1/4 HP, 115/230V, 1 Ph	$1,463
MI 10	1000	1/4 HP, 115/230V, 1 Ph	1,735
MI 12	1200	1/3 HP, 115/230V, 1 Ph	1,876
MI 16	1600	1/3 HP, 115/230V, 1 Ph	2,153
MI 20	2000	1/2 HP, 115/230V, 1 Ph	2,425
MI 24	2400	1/2 HP, 115/230V, 1 Ph	2,691
MI 40-1	4000	1 HP, 220/460V, 3 Ph	4,672
MI 48-1	4800	1 HP, 220/460V, 3 Ph	4,912

Portable Model

| MI 10P | 1000 | 3/4 HP, 115V, 1 Ph | 3,031 |
| MI 10P2 (2 pass) | 1000 | 3/4 HP, 115V, 1 Ph | 4,024 |

8 Spare Cell	800		366
10 Spare Cell	1000		413
12 Spare Cell	1200		455

Added to 115/230V, 1 Ph Motor Price for 3 Ph Motor   157

Available Accessories

Remote Power Pack		131
Water Wash Spray Header - Price per cfm		.19
CONTROL A	A manually initiated timer designed to wash and dry the Precipitator through a single push button control (includes solenoid valve)	335
CONTROL A-1	Used with detergent pump	379
INITIATOR CLOCK	Used with Controls A and A-1 to provide a completely automatic wash and dry cycle of the Precipitator at a selected hour daily or weekly.	121
DETERGENT PUMP	Includes five gallons of detergent	605
SOURCE CAPTURE PLENUM	(Price upon application)	
CHARCOAL FILTER	(Price upon application)	
HIGH VOLTAGE CABLE, per foot		.38

TERMS: All prices are F.O.B. factory, Monroeville, Pa.
        Prices and specifications are subject to change
        without notice.

UNIVERSAL AIR PRECIPITATOR CORPORATION
1500 McCully Road, Monroeville, PA. 15146
(412) 372-0706
1-800-255-8406

Figure 3.2 Sample vendor price list (courtesy Universal Air Precipitator Corporation).

Figure 3.2 is an example. Price lists like this one are invaluable, for they often contain information about auxiliary components, different models, and other equipment the estimator has not even conceived of. (Indeed, these "extras" may not even have been mentioned in the vendor's literature.)

Quotations usually will address either *packaged* or *custom* control equipment. *Packaged* equipment consists of "off-the-shelf" units, the size, materials, and configurations of which fit the source to be controlled. Typically, they are denoted by model numbers ("XYZ-123") in the vendor's literature. Although vendor literature rarely contains the costs for these models, it often includes useful design and operating data, such as the dimensions of the unit and the pressure drop. These packaged units are generally self-contained — prepiped, prewired, skid-mounted — and ready for immediate hook-up at the source. All the customer has to supply is the foundation, ducting (from the emission source to the control device), and the utilities: power, water, steam, compressed air, etc. Consequently, the costs required for installing packaged units are relatively low — 20 to 50% of the equipment price. Moreover, packaged units usually come with full instrumentation, fans, pumps, and even a short ("stub") stack. This keeps down the cost of auxiliary equipment. But the customer may need to augment these items with additional equipment, such as a "booster" fan to overcome pressure losses upstream of the control device. (See Chapter 4 for more information on auxiliary equipment.)

One drawback of packaged units is that they come in discrete capacities only. These capacities may or may not correspond to the size of the emission source, so that the customer may have to pay for excess capacity to be able to use a packaged unit. Secondly, packaged units are limited to "normal" applications — moderate flowrates, temperatures, moisture contents, etc. When a packaged system seems inappropriate, a vendor may quote a price for a *custom* control device, one whose design will conform precisely to the source specifications. As we might expect, a custom unit will almost always cost more than a packaged unit of the same size, because it requires more engineering, more site-specific input data, perhaps special materials of construction, and more sophisticated instrumentation.

Some vendors fabricate both packaged and custom units, while others manufacture either. Further, because of their large size, some items — such as tall stacks for dispersing large quantities of emissions over huge areas — are invariably custom-built. This introduces another way to distinguish equipment: *shop-fabricated* versus *field-erected*. The former items are built in the vendor's shops and shipped to the job site, where they are connected to the other equipment (if custom units) or simply hooked up (packaged units). Conversely, field-erected units are not only connected but are actually *built* onsite. Large ESPs typify field-erected controls. For several reasons — the higher cost of field versus shop labor, mainly — field-erected systems usually are more expensive than shop-fabricated. Thus, field-erection is avoided unless absolutely required. Again, size generally distinguishes between the two equipment categories. That is, the size of shop-fabricated equipment is limited by shipping constraints (e.g., the diameter of vessels that may be shipped by rail has been limited to 12 feet).

Whichever category of control device best suits the source, one should solicit several quotations for it. But *how many*? Unless only one vendor can supply the type of control device needed, at least two—preferably three—quotes should be requested. If nothing else, this lets us know the approximate spread in costs between manufacturers of the same device. This distribution also serves to highlight design differences between vendors' products—differences that the prices reflect. Not only are there design differences between different control device categories—say, a venturi versus a wet impingement scrubber— there are design variations among different manufacturers of the same product. This is so even though the basic collection/conversion mechanism may be the same for all such devices. These variations may include the use of more sophisticated instrumentation, novel techniques for pretreating the inlet waste gas, and more durable materials of construction.

Once all the quotes are in hand, the cost data must be analyzed and correlated against the equipment size parameter(s). Below we'll discuss techniques for doing this.

Owners and Operators of Control Systems

Potential sources of cost information also include firms who have installed air pollution control systems. It is difficult to question real costs that have been, or are being, incurred. Unlike estimates, which are inherently uncertain, actual cost data are *certain*. But there are some problems with using "real world" cost data. These include:

- *inconsistent specifications*, wherein the specifications for the firm differ significantly from those of the source in question. These are not only differences in such parameters as the waste gas flowrate, but include discrepancies in the basic design features of the system, such as the type and number of equipment items included.

- *different cost bases*, such as the vintage of the costs (e.g., 1982 versus 1988 dollars) or the unit costs for utilities, labor, and operating materials.

- *hidden "padding"*: Unfortunately, there are firms who purposely inflate their control system costs. They do so by including "spare" equipment (e.g., fans) in the design or, more insidiously, including inflated or superfluous costs in the estimate. (The contingency is a favorite cost to pad.) Rooting out and removing this "gold plating" may not be easy—or even possible, unless the estimator can obtain the full set of specifications and detailed capital and annual costs for the firm's control system. Such details are rarely available to outsiders. However, if the estimator is an employee of the firm, he or she likely can obtain these details and make appropriate adjustments.

Given enough information, one can adjust a firm's estimate to put it on the same basis as the source's. But if these adjustments are significant—30% or more to a single cost or technical parameter—so much error may be introduced that the firm's cost may be valueless. Moreover, making these adjustments may require a great deal of time and effort. These and other reasons may obviate the advantages of using "actual" costs—i.e., certainty and convenience.

Technical Publications

For those who don't mind digging, cost data may also be found in books, reports, journals, and (rarely) in vendor's technical bulletins. Very few books are devoted to—or even feature—air pollution control costs. (This one excepted, of course!) Fairly complete cost data for process equipment (fans, vessels, heat exchangers, and the like) do appear in such books as Peters and Timmerhaus' *Plant Design and Economics for Chemical Engineers*.[3]

As for journals, many engineering magazines publish articles *containing* cost data, though precious few run pieces *devoted to* cost matters. *Chemical Engineering* is a notable exception. Its "Cost File" department (ably edited for many years by Jay Matley) regularly publishes articles on process equipment-related sizing and costing. Some of these articles have even addressed pollution control costs. (For example, two article series—one on air pollution control equipment, the other on hazardous waste incinerators—have appeared in the "Cost File" since 1980.)

The *Journal of the Air Pollution Control Association* (*JAPCA*) is another source of air pollution control costs. Recent articles on the sizing and costing of incinerators, fabric filters, and electrostatic precipitators have appeared in *JAPCA*. In addition, *JAPCA* has run pieces on other cost topics, such as the results of a power plant FGD retrofit study.

Reports—mainly those written by or for the U.S. Environmental Protection Agency—provide another font of cost information. These include *Capital and Operating Costs of Selected Air Pollution Control Systems* and its sequel, the *EAB* [Economic Analysis Branch] *Control Cost Manual*.[4,5] Both reports focus on "add-on" control devices and auxiliaries. The former is fairly comprehensive; however, its costs are old (1977 dollars). The latter, though containing more current cost data (1986 or newer), addresses (as of this writing) only a limited number of control devices. Both reports are available through the National Technical Information Service (NTIS).

The Electric Power Research Institute (EPRI) and the Tennessee Valley Authority (TVA) also publish cost reports on air pollution control-related subjects. However, most focus on controlling emissions from power plants (steam-electric and other). Accordingly, the EPRI and TVA reports lean heavily toward FGDs and such process-related controls as boiler combustion modifications to reduce nitrogen oxide emissions. (See reference 6, for instance.)

Finally, vendor catalogs and price lists can provide very valuable cost data. But for reasons stated earlier, few vendors make this information available to the general public. Manufacturers of instrumentation, laboratory and safety equipment, and small auxiliaries (e.g., pumps) are exceptions. Upon request, these vendors will send price lists, or simply include them with the colorful literature they disseminate. Like any other cost data, though, these lists become obsolete periodically and should be used with this in mind.

ANALYZING COST DATA

Wherever you get them from, raw cost data must be adjusted (if necessary) and correlated before they will be of much use. Ideally, all cost data obtained for making capital and annual cost estimates will be of the same vintage. If not, some of the costs will have to be adjusted — that is, updated (or back-dated) to some common vintage or reference date. This is done via an *escalation index*, viz.:

$$\text{Cost (new)} = \text{Cost (old)} \times \text{Index (new)}/\text{Index (old)} \qquad (3.1)$$

where Index (new) and Index (old) refer to the reference date and the base (original) date of the raw costs.

As Chapter 8 discusses in much more detail, several different published indices may be used for air pollution control equipment. The choice of index will depend on the kind of equipment being costed. But, as a rule of thumb, *costs older than five years should not be escalated*, because equipment price increases rarely "track" very closely with published indices. This is especially so in periods of high inflation, when even the five-year limit may be too long. (Conversely, when inflation is atypically low, the rule may be stretched a bit.)

Correlation is the next step. Here's another rule to remember: do *not* include in a *single* correlation (e.g., equipment price vs. waste gas flowrate) costs of *different* base dates, *even if* all costs have been escalated to the reference date. To do so would only serve to dilute the quality of the newer data. The only exceptions to this rule would be if there were a serious gap in the data set — a gap which could not be filled without including costs of different base dates.

Of course, when making a *different* correlation (say, electric power cost vs. gas flowrate) the estimator may have to use costs of a different *base* date from those used in another correlation. But as these are different *kinds* of costs, there is no danger of diluting the other data. (Needless to say, *all* costs must be at, or adjusted to, the reference date for the estimate.)

Correlating cost data typically involves plotting costs against some variable or variables. Consider the hypothetical equipment costs for packaged thermal incinerators shown in Table 3.2. In this data set, equipment cost is the *dependent* variable, while gas flowrate and heat recovery are the *independent* variables. One could plot cost versus flowrate, with heat recovery as a parameter, and obtain a family of curves. Alternatively, cost could be plotted against heat recovery, with flowrate as a parameter. The plotting could be done on arithmetic, logarithmic-logarithmic, or other coordinates. As it turns out, these data form straight lines on log-log paper. The equations for the lines are:

Table 3.2 Hypothetical Equipment Costs for Packaged Thermal Incinerators

Inlet Flow (scfm)	Heat Recovery (%)	Equipment Cost ($)
10,000	35	85,000
	50	95,000
20,000	35	120,000
	50	134,000
50,000	35	190,000
	50	212,000

$$\text{Cost (35\% heat recovery)} = 850Q^{0.5} \tag{3.2}$$

$$\text{Cost (50\% heat recovery)} = 950Q^{0.5} \tag{3.3}$$

where Q = inlet gas flowrate (scfm).

Finally, a more mathematically inclined estimator might want to develop a multi-variable correlation, weaving together cost, flowrate, and heat recovery. But because in this case there are only data for two heat recoveries, 35% and 50%, such a correlation might be of limited value, especially since 35% and 50% are so close together. Regardless of whether you select a single- or a multi-variable correlation, be sure that the correlation would be mathematically *continuous* (or nearly so) over the range of the independent variable. That is, if packaged incinerators were only sold with heat exchangers that obtained only *discrete* heat recoveries (say, 35 and 50%), a *continuous* correlation between price and heat recovery would be completely unrealistic.

The foregoing illustrated how to treat cost data from a single vendor. What if we had obtained incinerator costs for these flowrates and heat recoveries from several vendors? One approach would be to *average* the vendor prices for each flowrate-heat recovery set and then plot these averages against flowrate (or heat recovery). By so doing, we would be "smoothing out" design, pricing, and other differences for the vendors who supplied the costs. If these design/pricing differences were significant, however, this averaging approach might not realistically portray the data.

When we obtain data from several equally reliable sources, it might be better to use *regression analysis*. We might regress the data against a linear, quadratic, log-log, semi-log, or other equational form, whichever offers the best fit. In this way, we would treat each cost datum as an *independent* observation. The techniques of regression analysis are used to measure the "goodness of fit" of data against a certain type of correlation. In general, the higher the "degrees of freedom" (number of data points minus 2), the better the correlation, as measured by the familiar "r" correlation coefficient. According to Miller,[7] the closer "r" is to unity, the greater the probability that a real correlation between two variables exists.

DISPLAYING COST DATA

The three main forms for displaying cost data are:

- graphs or figures
- tables
- equations

Regardless of which form is used, certain basic information should be given, viz.:

- equipment specifications (what prices include, materials of construction, etc.)
- reference and base (original) dates for cost data
- source(s) of the cost data
- accuracy of the cost data
- whether extrapolation is permitted

Figure 3.3 is a hypothetical cost curve, in which the price of shaker bag-houses has been plotted against the cloth (bag) area (in sq ft). Note that the price includes a pressure-type, basic baghouse (without bags), based on carbon steel construction. The base date for these prices, obtained from two fictional vendors, was July 1985 and the reference date (i.e., date to which they were escalated) is July 1988. The figure also shows that the costs should *not* be extrapolated beyond the range of the correlation – a least-squares fit, as the correlation coefficient (r) indicates.

Figure 3.3 is, of course, a graphical data display. This and the other two forms have their pros and cons. Figures have the visual edge over the others, in that they allow us to actually see the functional relationships among the dependent and independent variables. This form is especially useful in cases where cost is a function of several variables, such as in the above incinerator illustration. But figures are often hard to read, especially those plotted on log-log coordinates, where interpolation between "hash marks" can be difficult and error-inducing.

Tables offer the advantage of displaying large amounts of data in relatively little space. Furthermore, unlike figures, tables show exactly the cost data corresponding to a given set of variables. But *interpolating* between the fixed points on a table can be treacherous. Consider the incinerator costs given in Table 3.2. Suppose we needed the cost of a 30,000-scfm unit with 50% heat recovery. Without knowing anything about the relationship between cost, flowrate, and heat recovery, we likely would interpolate linearly between the costs for the 20,000 and 50,000 scfm units, viz.:

$$\text{Cost (\$1000)} = \$120 + ([\$212 - \$134]/[50 - 20]) \times (50 - 20)$$
$$= \$146$$

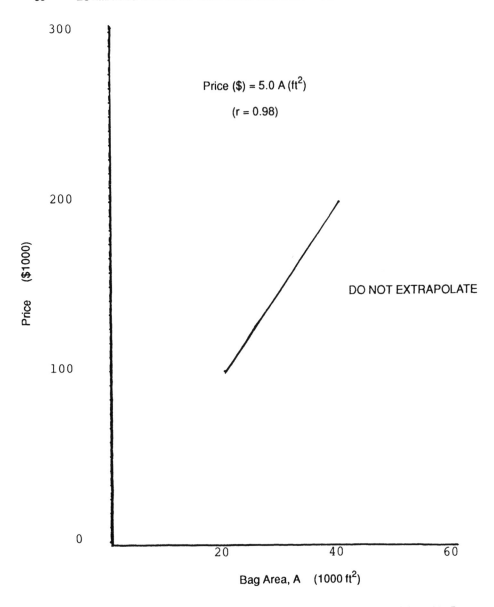

Price (\$) = 5.0 A (ft^2)

(r = 0.98)

DO NOT EXTRAPOLATE

Price   (\$1000)

Bag Area, A   (1000 ft^2)

Figure 3.3 Hypothetical fabric filter cost curve. *Reference* date for prices: July 1988. *Base* date: July 1985. Prices (±20% accurate) are for pressure-type shaker bag-houses, carbon steel construction, *without* bags. *Sources*: Purer Air, Inc. (Hermitage, PA); Fabri-Clean Corp. (Masury, OH).

Compare this to the cost obtained from Equation 3.3: \$165,000. This represents a large difference by any measure.

These problems – cumbersome interpolations and curve-reading errors – are avoided with *equations*. Equational data are more concise, more rigorous, and much easier to computerize than either graphical or tabular data. In addition, equations are easy to adjust, if needed. This is particularly advantageous with cost data that often must be updated via escalation index multipliers. The only drawback with an equation is that it provides no "picture" of the functional relationship. And there are occasions (e.g., regression fits) when such a picture is desirable.

Finally, a few words about *units*. Note that *English* units have been used exclusively in this book. Being a "mature" engineer, this author is more comfortable with the English than with metric or other unit systems. More importantly, English units are used herein because the variables against which cost data are correlated in the literature are typically expressed in them. Also, both equipment vendors and the American Association of Cost Engineers seem to favor them. All of this reflects the general preference for English units in the cost engineering profession.

RETROFIT COSTS

This topic may seem to have escaped from Chapter 2, which discussed the cost estimating procedures used in this book. These procedures, however, were mainly geared to estimating costs for controlling emission sources in *new* ("grass roots" or "green field") facilities. Conversely, retrofit control costs pertain to *existing* sources. The distinction is more than academic. In fact, retrofit costing is virtually a subject unto itself. This section will present techniques for estimating retrofit costs. But before doing that, we need to qualitatively discuss which *kinds* of capital and annual costs are (and are not) affected in a retrofit situation.

First, consider the *control device* itself. Typically, the size (and cost) of the control device would not be significantly different in a retrofit installation. This is because the waste gas stream parameters and other inputs that determine the price of the device probably would not be different if the source were located in an existing rather than a new facility.

The same cannot necessarily be said for the *auxiliary equipment*, however. Most notable is the ductwork, for in many retrofits rather long duct runs (several hundred feet in some cases) are needed to connect the control device to the existing emission source. Along with longer runs, retrofit ductwork may require extra dampers, tees, elbows, and other expensive fittings. When designing a new facility, the control device can be situated much closer to the source to avoid these expensive ductwork costs.

In retrofit situations, several direct and indirect installation costs may also increase significantly, including:

- *Handling and erection.* Because the space available for the control system may be limited in an existing plant, special care may need to be taken when unloading, transporting, and placing the equipment.

- *Piping, insulation, and painting.* These costs could increase significantly in a retrofit situation. Like ductwork, large amounts of piping may be needed to "tie in" the control device; of course, the more piping and ductwork required, the more insulation and painting needed.

- *Site preparation.* Unlike the other categories, this cost may actually decrease, for most of this work would have been done when the original facility was built.

- *Facilities and buildings.* Potentially, retrofit costs here could be the largest for this category, especially so for the facilities. For example, if the control system requires a great deal of power for operation (e.g., a venturi scrubber), the facility's power plant may not be able to service it. In such a case, the facility would either have to purchase additional electricity from a public utility, expand its power plant, or build another one. In any case, the cost of electricity supplied to that venturi would surely be higher than if the scrubber had been installed in a new plant where adequate provision for its electrical needs could have been made. And if space limitations force installation of the control system outdoors, a *building* may need to be erected to protect it from the elements. In a new plant, conversely, this same system could have been placed indoors.

- *Engineering and supervision.* Designing a control system to fit into an existing plant normally requires extra engineering, especially when the system is extraordinarily large, heavy, or utility-consumptive. For the same reasons, extra supervision may be needed when the installation work is being done.

- *Lost production.* If the control system cannot be tied into the process during normally scheduled maintenance periods, part or all of the process may have to be shut down. The *net* (after-tax) value of the revenue not earned during this time is a bonafide retrofit expense.

- *Contingencies.* Due to the uncertain nature of retrofit estimates — even those based on "hard" data — the estimator probably should increase the contingency or uncertainty factor.

These points address some of the costs that may increase in retrofit situations. However, there may be other cases where the retrofitted installation costs would be *less* than the cost of installing the system in a new plant. This could occur, for instance, when one control system, say an ESP, is being replaced by another — a baghouse, say. In such a case, the ductwork, stack, offsite facilities, etc., for the ESP could be adequate for the replacement system.

Table 3.3 Retrofit vs. New Facility Capital Costs for Hypothetical Carbon Adsorber System

Cost Item	Cost ($1,000)		Retrofit/New (ratio)
	New	Retrofit	
Direct Costs			
Purchased equipment (PE):			
Control device	150	150	1
Auxiliaries			
Ductwork	40	80	2.0
Stack	10	10	1
Instrumentation	20	24	1.2
Taxes	6	7.2	1.2
Freight	10	12	1.2
Total PE	236.0	283.2	1.2
Installation:			
Foundations, supports	18.9	34.0	1.8
Handling, erection	33.0	79.3	2.4
Electrical	9.4	17.0	1.8
Piping	4.7	11.3	2.4
Insulation	2.4	5.7	2.4
Painting	2.4	5.7	2.4
Site preparation	25.0	0	0
Facilities and buildings	45.0	65.0	1.4
Total Direct Installation	140.8	218.0	1.5
Total Directs	376.8	501.2	1.3
Indirect Costs			
Installation:			
Engineering, supervision	23.6	56.6	2.4
Construction, field	11.8	22.7	1.9
Construction fee	23.6	42.5	1.8
Startup	4.7	5.7	1.2
Performance test	2.4	2.8	1.2
Contingencies	7.1	28.3	4.0
Total Indirects	73.2	158.6	2.2
TOTAL CAPITAL INVESTMENT:	450	660	1.5

How can we quantify retrofit cost increases? First, because each retrofit situation is different, no general rule can be developed, such as: "retrofit installation cost equals twice the new facility installation cost." However, we can use an approximate, semi-quantitative procedure based on applying the adjustment factors in Table 2.3 to the installation factors in Table 2.2. For instance, we could adjust the "handling and erection" factor by assuming that it would double in a retrofit situation. The "site preparation," "facilities and buildings," "engineering and supervision," "construction and field," "construction fee," and "contingency" installation factors could also be increased (or decreased) in making a retrofit cost estimate.

This adjustment process would yield a *retrofit* installation factor, which one could use in a hypothetical retrofit situation. Table 3.3 shows these adjusted factors for a hypothetical carbon adsorber system. Note that the retrofit total capital investment would be approximately 50% higher than the TCI for the same adsorber installed in a new facility. Keep in mind that these results are approximate, based on "guesstimates," nothing more. Moreover, these results are meant for illustrative purposes only and should not be taken literally.

A similar approach for estimating retrofit costs is provided by the Electric Power Research Institute.[6] EPRI developed a set of guidelines for estimating the capital and operating and maintenance (O&M) costs for six types of flue gas desulfurization systems at existing steam-electric power plants. The guidelines were to enable utilities to convert EPRI (or other reliable) new plant costs to retrofit costs for FGDs installed at *specific* (not generic) power plants. The guidelines addressed changes in:

- project scope (e.g., addition/modification of chimneys)
- process (waste gas flow per megawatt, coal sulfur content, etc.)
- location (climate, material and labor cost indices, etc.)
- retrofit adjustments (access and congestion, ductwork tie-in difficulty, etc.)

To illustrate the sensitivity of the capital cost to changes in these parameters, EPRI developed retrofit costs for four "case studies." The base capital costs for these studies varied from 1.2 to 1.6 times the new plant costs. (Compare this range with the retrofit/new plant cost ratio in Table 3.3.)

Of the parameters EPRI found that influenced the installed capital cost, those producing the largest cost increases were the material and labor cost index (up to 30%) and the retrofit adjustments (5 to 70%). The biggest contributors to the retrofit adjustments were the accessibility, congestion, and ductwork distance. This tends to confirm the retrofit discussion above.

Interestingly, the EPRI procedure did *not* adjust any of the indirect capital costs (e.g., engineering), unlike the procedure illustrated in Table 3.3. And although the EPRI retrofit procedure is detailed, well-documented, and straightforward to use, it is limited to power plant FGDs. Still, it offers enough suitable "raw material" for anyone interested in applying their procedure to other retrofit situations.

*   *   *

In this and the previous chapter, we have covered the elements comprising cost estimates, how to estimate these costs, and a methodology for sizing and costing control methods. In following chapters, we'll go from the general to the specific, by *applying* these procedures to the sizing and costing of a variety of control equipment. These chapters will also present current equipment costs correlated against appropriate sizing parameters, and where appropriate, factors for estimating installation, maintenance, utilities, and other costs. We'll begin in Chapter 4 with auxiliary equipment—fans, ductwork, stacks, and other equipment that are necessary to the operation of control systems.

REFERENCES

1. *Chemical Engineering*, April 11, 1988, p. 15.
2. *Pollution Equipment News: 1988 Buyer's Guide.* Pittsburgh: Rimbach Publishing, November 1987.
3. Peters, M. S., and K. D. Timmerhaus. *Plant Design and Economics for Chemical Engineers*, 3rd ed. New York: McGraw-Hill, 1980.
4. *Capital and Operating Costs of Selected Air Pollution Control Systems.* Research Triangle Park, NC: U.S. Environmental Protection Agency (1978) (NTIS PB-80-157282).
5. *EAB Control Cost Manual.* Research Triangle Park, NC: U.S. Environmental Protection Agency (1987) (NTIS PB-87-166583/AS).
6. Shattuck, D. M., et al. *Retrofit FGD Cost-Estimating Guidelines.* Palo Alto, CA: Electric Power Research Institute (CS-3696, Research Project 1610-1, October 1984).
7. Miller, R. E. "CE Tutorial—Statistics, Part 7: Correlation and Regression," *Chemical Engineering*, September 30, 1985, pp. 71-75.

CHAPTER 4

"Add-on" Controls I: Auxiliary Equipment

> This most excellent canopy, the air.
>
> *Hamlet* — William Shakespeare

In this and the next two chapters, we'll focus on "add-on" control devices and the auxiliary equipment that supports their operation. As explained in Chapter 3, "add-ons" are those devices that are installed downstream of stack emission sources to control particulate and/or gaseous emissions. Thus, they are "added on" to the source being controlled. Chapter 3 also listed some of the auxiliary equipment for these "add-ons." Many of these same items can also be used with non-add-ons, as well (e.g., fans with house ventilation systems for radon abatement). This chapter presents descriptions, sizing procedures, and equipment costs for the more commonly used auxiliaries. In order of appearance, the players are:

- fans and motors
- ductwork
- stacks
- cyclones
- screw conveyors and rotary air locks
- hoods
- buildings

FANS

Many air pollution control devices come equipped with built-in fans powerful enough to convey the waste gas to and from them. In those cases where a fan is not provided or the one provided is undersized, one must install a separate fan in the control system.

Description

There are two main categories of fans: (1) *centrifugal (radial-flow)* in which the gas flows perpendicularly to the rotor axis of rotation and (2) *axial-flow (propellor)*, wherein the gas flows parallel to the rotation axis. Because the axial-type fans are rarely used in air pollution control systems, we'll limit the discussion to centrifugal fans.

As its name implies, a centrifugal fan operates by imparting centrifugal force to an element of fluid (gas). The fan consists of a wheel (rotor) mounted on a shaft rotating in a scroll-shaped housing. After entering at the rotor's eye, the gas makes a right-hand turn and is shoved by centrifugal force through the rotor blades into the housing. This force imparts *static pressure* (SP) to the gas, while the scroll converts part of the *velocity pressure* (VP) into static pressure. The design of the rotor blade distinguishes the three types of centrifugal fans made. Of these three, two types are most commonly used in control systems:[1]

- *Backward-curved blade:* In these fans, the blades are inclined opposite to the direction of rotation. The 14 to 24 blades are supported by a solid steel backplate and shroud ring. The operating efficiency of backward-curved blade fans is relatively high, but so is their price. Moreover, their blades are not designed to handle large concentrations of dusts and fumes. For this reason, they are always installed on the "clean" (downstream or suction) side of the control device.

- *Straight-blade:* Also known as "radial-tip," these fans have 5 to 12 straight blades which may be attached to the large-diameter rotor by a solid steel backplate or a spider built up from the hub. These fans are used to convey waste gases with high concentrations of particulate matter and fumes. Different blade and scroll designs have been developed to handle specific types of contaminants. In severe service, blades are fabricated of abrasive-resistant alloys or covered with rubber. Finally, though cheaper than the backward-curved, straight-blade fans are less efficient.

Both of these fan types are sold either individually or with motors and drives ("packaged fans"). In either case, the motor is connected to the fan directly ("direct-driven") or via a V-belt ("belt-driven").[2] Direct-driven fans offer a more compact motor-fan assembly, while ensuring a constant fan speed. Also, they are not plagued by the belt slippage that occurs in poorly maintained belt-driven fans. However, belt-driven fans allow the user to vary the fan speed over a wide range by simply changing the belts. With direct-driven fans, the speeds are limited only to those the motor can attain.

Sizing Procedure

Fan manufacturers customarily provide a series of fans that are characterized by constant ratios of linear dimensions and constant angles between fan parts. These are known as *homologous series* fans—fans that are identical in all respects, save for size. The primary sizing parameter for fans is the *wheel diameter*. Manufacturers typically use the wheel diameter as the basis for building fans in a homologous series. For example, the wheel diameter in Model A might be 15 in.; in Model B, 18 in.; in Model C, 21.5 in. In this case, the ratio between the diameters is a constant 1.2.

In air pollution control systems, one selects or designs a fan to overcome the *system resistance* or *pressure drop* between the emission source and the stack outlet. To overcome this loss and to convey the gas through the system at the desired velocities, the fan generates a *total pressure* (TP) which is the sum of the static and velocity pressures, the latter being associated with the moving gas. Equationally:

$$TP = SP + VP \qquad (4.1)$$

Both the total and static pressures vary as functions of the gas volume passing through the fan. At zero gas flowrate, VP = 0 and TP = SP. Conversely, at maximum design flowrate, SP = 0 and TP = VP. The functional relationships between these fan pressures and flowrates are shown on fan *characteristic curves*, which fan manufacturers construct, based on experimental measurements. Characteristic curves also plot the fan "mechanical efficiency" — the ratio of horsepower output to the horsepower input (or *brake horsepower*, BHP). The maximum fan efficiency calculated based on the total fan pressure (TP) typically ranges from 50 to 70%. The required input or brake horsepower is calculated as follows:

$$BHP = 0.0001575Q\Delta P/n' \qquad (4.2)$$

where   Q = gas volumetric flowrate (actual cubic feet/min)
$\quad\quad\quad\quad$ ΔP = total system pressure loss to overcome (inches of water, column)
$\quad\quad\quad\quad$ n' = fan efficiency (fractional)

(Note: "n'," the fan efficiency, is NOT the same parameter as "n," the combined fan-motor efficiency presented in Chapter 2.)

A fan is selected or designed to meet the flowrate and pressure drop needs of a control system while operating at its maximum mechanical efficiency. The maximum efficiency is typically reached between 1/3 and 2/3 of the maximum design flowrate. Generally, the flow to and from the fan is controlled by *inlet and outlet dampers*. Certain relationships ("Fan Laws") govern the operation of fans in a homologous series. These relationships correlate two or more of the following independent variables:

- wheel diameter (D)
- fan speed (N)
- gas density (r)
- system resistance (ΔP)

- brake horsepower (BHP)

For the special case of fans operating at the same "point of rating," the following relationships hold:

$$Q = k_1(D^3)N \tag{4.3}$$

$$\Delta P = k_2(D^2)N \tag{4.4}$$

$$BHP = k_3(D^5)(N^3)r \tag{4.5}$$

where the k's are constants.

For example, a fan with twice the wheel (impeller) diameter could convey eight times the volumetric flowrate at the same speed.

Remember that these equations are *only* valid for fans belonging to a homologous series. For non-homologous fans, the performance specifications should be obtained from the manufacturer's *multirating tables*. In these tables, values of static pressure are arrayed against volumetric flowrates. For each flowrate-pressure pair, the table also lists the fan speed, outlet velocity, and brake horsepower required to deliver that flowrate and pressure. Figure 4.1 is a multirating table for three radial blade centrifugal fans. Note that the three fan models shown span an operating flowrate of approximately 3,000 to 27,000 standard ft³/min (scfm) over a static pressure range of 2 to 22 inches of water. But for optimum mechanical efficiency, we should limit our selection to the values within the area enclosed by the solid lines.

Also note that the flowrates and pressures are measured at standard conditions—typically 70°F and 14.696 lb/in², absolute (psia), or "sea level installation." At most installations, the conditions will be nonstandard. When selecting a fan in these cases, we must first adjust the static pressure measured at the installation conditions. This formula provides such an adjustment:[3]

$$P_i/P_s = 530/t[\exp(-00003.53z')] \tag{4.6}$$

where P_i, P_s = installation and standard pressures, respectively
t = absolute temperature at installation (R)
z' = installation altitude above sea level (ft)

Example: A 5-inch static pressure drop must be overcome in a control system installed at a 1500-ft. elevation. The temperature at the fan is 200°F. What is the equivalent static pressure at standard conditions (sea level and 70°F)? Substituting this temperature and altitude into Equation 4.6, we obtain:

$$P_i/P_s = 0.847$$

or:

$$P_s = P_i/0.847 = 5.9 \text{ in.}$$

Therefore, we would select a fan capable of overcoming a 5.9-in. static pressure drop.

404 LS

Inlet diameter: 23" O.D.
Outlet area: 2.90 sq. ft. inside
Wheel diameter: 40"
Wheel circumference: 10.47 ft.

CFM	OV	2"SP RPM	2"SP BHP	4"SP RPM	4"SP BHP	6"SP RPM	6"SP BHP	8"SP RPM	8"SP BHP	10"SP RPM	10"SP BHP	12"SP RPM	12"SP BHP	14"SP RPM	14"SP BHP	16"SP RPM	16"SP BHP	18"SP RPM	18"SP BHP	20"SP RPM	20"SP BHP	22"SP RPM	22"SP BHP
2900	1000	476	1.60	666	3.49	813	5.66	939	8.09	1050	10.7	1150	13.5	1243	16.5	1328	19.6	1408	22.9	1485	26.3	1559	29.9
3480	1200	481	1.85	668	3.91	814	6.25	939	8.85	1049	11.6	1148	14.6	1240	17.7	1326	21.0	1407	24.5	1483	28.0	1556	31.8
4060	1400	489	2.15	672	4.38	816	6.89	940	9.64	1049	12.6	1149	15.7	1240	19.0	1325	22.4	1405	26.0	1482	29.8	1553	33.6
4640	1600	497	2.47	678	4.89	820	7.57	942	10.5	1051	13.6	1150	16.9	1241	20.3	1325	23.9	1405	27.7	1480	31.5	1552	35.5
5220	1800	508	2.82	684	5.43	825	8.29	945	11.3	1053	14.6	1151	18.1	1241	21.7	1326	25.5	1404	29.3	1479	33.3	1552	37.6
5800	2000	520	3.23	692	6.03	830	9.04	949	12.3	1056	15.7	1153	19.3	1242	23.1	1327	27.0	1405	31.0	1481	35.3	1550	39.5
6380	2200	531	3.65	700	6.66	836	9.86	955	13.3	1061	16.9	1157	20.7	1244	24.5	1328	28.6	1407	32.9	1481	37.2	1552	41.7
6960	2400	544	4.13	710	7.34	845	10.8	960	14.3	1066	18.1	1161	22.1	1249	26.1	1331	30.3	1409	34.7	1483	39.2	1553	43.8
7540	2600	558	4.65	720	8.07	853	11.7	968	15.5	1071	19.4	1165	23.5	1252	27.7	1333	32.1	1411	36.6	1485	41.1	1555	46.1
8700	3000	588	5.84	744	9.70	873	13.8	985	17.9	1085	22.2	1177	26.6	1263	31.3	1343	35.9	1420	40.9	1491	45.8	1562	50.9
9860	3400	620	7.24	770	11.6	895	16.0	1004	20.6	1102	25.3	1192	30.1	1276	35.0	1356	40.2	1431	45.4	1501	50.6	1570	56.1
11020	3800	654	8.94	797	13.7	918	18.5	1024	23.5	1121	28.7	1209	33.8	1292	39.2	1369	44.6	1443	50.1	1513	55.8	1579	61.5
12180	4200	691	10.9	827	16.0	944	21.4	1047	26.8	1142	32.3	1230	38.0	1309	43.6	1386	49.6	1459	55.4	1527	61.3	1594	67.5
13340	4600	729	13.2	859	18.8	972	24.5	1073	30.4	1166	36.4	1251	42.4	1330	48.5	1404	54.7	1476	61.0	1543	67.3	1608	73.8
14500	5000	768	15.9	892	21.9	1002	28.0	1100	34.3	1190	40.7	1273	47.2	1351	53.6	1424	60.2	1494	67.0	1561	73.7	1624	80.4
15660	5400	809	19.0	928	25.3	1033	31.9	1128	38.6	1216	45.4	1297	52.3	1373	59.2	1445	66.2	1516	73.4	1580	80.4		
16820	5800	851	22.5	964	29.2	1065	36.2	1157	43.3	1244	50.6	1323	57.9	1399	65.3	1469	72.6	1536	80.1	1601	87.7		

454 LS

Inlet diameter: 26" O.D.
Outlet area: 3.69 sq. ft. inside
Wheel diameter: 45½"
Wheel circumference: 11.81 ft.

CFM	OV	2"SP RPM	2"SP BHP	4"SP RPM	4"SP BHP	6"SP RPM	6"SP BHP	8"SP RPM	8"SP BHP	10"SP RPM	10"SP BHP	12"SP RPM	12"SP BHP	14"SP RPM	14"SP BHP	16"SP RPM	16"SP BHP	18"SP RPM	18"SP BHP	20"SP RPM	20"SP BHP	22"SP RPM	22"SP BHP
3690	1000	421	2.02	590	4.42	720	7.18	832	10.3	930	13.6	1019	17.2	1101	21.0	1177	24.9	1247	29.1	1316	33.4	1381	38.0
4428	1200	427	2.35	592	4.97	721	7.93	832	11.2	930	14.8	1017	18.5	1099	22.5	1176	26.7	1247	31.1	1315	35.6	1379	40.4
5166	1400	433	2.72	596	5.56	724	8.74	833	12.2	930	16.0	1018	20.0	1099	24.2	1174	28.5	1245	33.1	1313	37.8	1377	42.7
5904	1600	441	3.13	600	6.20	727	9.60	835	13.3	931	17.2	1019	21.4	1099	25.8	1174	30.4	1245	35.1	1312	40.1	1377	45.3
6642	1800	450	3.59	606	6.89	731	10.5	837	14.4	933	18.6	1020	22.9	1100	27.5	1175	32.4	1245	37.2	1313	42.5	1376	47.7
7380	2000	460	4.09	613	7.65	736	11.5	841	15.6	936	20.0	1022	24.6	1101	29.3	1176	34.4	1245	39.5	1313	44.9	1374	50.2
8118	2200	471	4.63	621	8.45	741	12.5	847	16.9	940	21.5	1025	26.3	1103	31.2	1177	36.4	1247	41.8	1313	47.2	1376	53.0
8856	2400	482	5.24	629	9.32	749	13.7	852	18.2	944	23.0	1029	28.0	1107	33.2	1179	38.6	1249	44.1	1314	49.9	1377	55.7
9594	2600	495	5.90	638	10.2	756	14.8	858	19.6	949	24.7	1032	29.9	1109	35.2	1182	40.8	1250	46.5	1317	52.5	1378	58.6
11070	3000	521	7.41	659	12.3	774	17.5	873	22.7	962	28.1	1043	33.8	1120	39.7	1190	45.7	1259	51.9	1322	58.2	1384	64.8
12546	3400	549	9.20	682	14.7	793	20.4	890	26.2	977	32.1	1056	38.2	1131	44.5	1202	51.0	1268	57.7	1331	64.4	1391	71.1
14022	3800	580	11.4	706	17.3	814	23.6	908	29.9	994	36.4	1072	43.1	1145	49.8	1214	56.7	1279	63.7	1341	70.9	1402	78.4
15498	4200	612	13.9	733	20.4	837	27.2	928	34.0	1012	41.1	1090	48.1	1160	55.4	1229	62.9	1293	70.4	1353	78.0	1413	85.8
16974	4600	646	16.8	761	23.9	861	31.2	951	38.6	1033	46.2	1108	53.9	1179	61.7	1244	69.5	1308	77.5	1369	85.7	1426	93.9
18450	5000	681	20.2	791	27.8	888	35.6	975	43.6	1055	51.8	1128	59.9	1197	68.2	1262	76.5	1325	85.1	1383	93.7	1440	102
19926	5400	717	24.1	822	32.2	915	40.6	999	49.1	1077	57.7	1149	66.2	1219	75.5	1283	84.4	1343	93.4	1400	102		
21402	5800	754	28.6	854	37.2	944	46.0	1025	55.0	1102	64.3	1173	73.6	1240	83.1	1302	92.3	1362	102	1419	111		

504 LS

Inlet diameter: 29" O.D.
Outlet area: 4.62 sq. ft. inside
Wheel diameter: 50½"
Wheel circumference: 13.22 ft.

CFM	OV	2"SP RPM	2"SP BHP	4"SP RPM	4"SP BHP	6"SP RPM	6"SP BHP	8"SP RPM	8"SP BHP	10"SP RPM	10"SP BHP	12"SP RPM	12"SP BHP	14"SP RPM	14"SP BHP	16"SP RPM	16"SP BHP	18"SP RPM	18"SP BHP	20"SP RPM	20"SP BHP	22"SP RPM	22"SP BHP
4620	1000	376	2.51	527	5.50	644	8.97	743	12.8	831	17.0	911	21.4	984	26.2	1051	31.1	1115	36.3	1176	41.8	1234	47.5
5544	1200	381	2.92	529	6.18	645	9.91	744	14.0	831	18.4	911	23.2	982	28.1	1050	33.3	1114	38.8	1174	44.5	1232	50.4
6468	1400	387	3.38	532	6.91	646	10.9	745	15.3	831	19.9	910	24.9	982	30.1	1049	35.5	1113	41.3	1173	47.3	1230	53.4
7392	1600	394	3.90	536	7.72	649	12.0	746	16.6	832	21.5	910	26.8	982	32.2	1049	37.9	1112	43.9	1172	50.1	1230	56.6
8316	1800	402	4.46	541	8.58	653	13.1	748	18.0	833	23.2	911	28.6	982	34.3	1050	40.4	1112	46.5	1173	53.1	1229	59.7
9240	2000	411	5.09	548	9.53	657	14.3	751	19.5	836	24.9	913	30.7	985	36.7	1051	42.9	1113	49.3	1173	56.1	1229	63.0
10164	2200	420	5.76	555	10.5	663	15.7	756	21.1	840	26.8	916	32.8	987	39.1	1052	45.5	1114	52.2	1173	59.0	1229	66.2
11088	2400	431	6.52	563	11.6	669	17.0	761	22.8	844	28.8	919	35.0	989	41.5	1054	48.2	1115	55.2	1174	62.3	1229	69.7
12012	2600	442	7.35	571	12.8	675	18.5	766	24.5	848	30.8	922	37.3	993	44.2	1058	51.2	1119	58.4	1176	65.7	1231	73.3
13860	3000	465	9.24	589	15.4	691	21.8	779	28.4	859	35.2	933	42.4	1000	49.6	1063	57.1	1124	64.9	1181	72.7	1237	81.0
15708	3400	491	11.5	609	18.3	708	25.4	795	32.7	872	40.1	944	47.8	1011	55.8	1074	63.8	1133	72.1	1189	80.4	1243	89.1
17556	3800	518	14.2	632	21.7	727	29.4	811	37.4	888	45.5	958	53.8	1023	62.2	1084	70.9	1142	79.7	1198	88.6	1252	98.0
19404	4200	547	17.3	655	25.5	748	34.0	831	42.7	904	51.3	974	60.4	1038	69.5	1098	78.7	1155	88.1	1210	97.7	1262	107
21252	4600	577	21.0	680	29.8	769	38.9	849	48.2	923	57.8	990	67.4	1053	77.1	1112	86.9	1168	96.9	1223	107	1274	117
23100	5000	608	25.2	706	34.7	793	44.5	871	54.5	942	64.7	1008	74.9	1069	85.3	1129	96.1	1183	106	1236	117	1286	128
24948	5400	640	30.1	734	40.2	818	50.7	893	61.3	962	72.2	1028	83.3	1089	94.4	1146	106	1200	117	1251	128	1302	140
26796	5800	673	35.8	763	46.4	843	57.5	916	68.8	985	80.4	1048	92.1	1108	104	1163	115	1217	127	1268	139		

Figure 4.1 Fan manufacturer's multirating table (courtesy L.R. Gorrell Co.).

Costing Procedure

Fans are sold with and without motors, in a variety of sizes, designs, and materials of construction. But all fans have one feature in common: their prices correlate best with their wheel (impeller) diameters. Why don't fan prices correlate with the flowrate, pressure drop, speed, or other parameters? Because every fan can operate over a range of flowrates and static pressures, so that a given fan can accommodate several control systems. Conversely, several *different* fans can handle a given stream. Figure 4.1 confirms this. Notice how one fan (model "454 LS") can deliver flows ranging from 3960 to 21,402 scfm at static pressures from 2 to 22 inches of water. However, as we said earlier, the area in Figure 4.1 bounded by the solid lines indicates the preferred operating range for model 454 LS. This "limits our options" somewhat. Even so, this fan can still handle a wide variety of streams.

But how do we select a fan that's "just right" for a given control system? Once we decide on the fan type (radial or backward-curved) and materials of construction, we would obtain the multirating tables for several suitable fans. From these tables we would select a fan (or fans) that would be able to convey our stream at close to maximum efficiency. If several such fans would do the job, the one to select would be the fan (and motor) having the lowest annual cost. (Procedures for calculating the total annual cost are provided in Chapter 2.)

The first (and key) step in this costing procedure is determining the price of the fan, motor, and supporting equipment (driver, starter, etc.).

Backward-Curved Fans

For *packaged* backward-curved centrifugal fans, the following equation provides prices (F.O.B. vendor) as a function of the fan wheel diameter (D, in.):[4]

$$Price (\$) = 42.3D^{1.20} \tag{4.7}$$

where
$$12.25 \leq D \leq 36.5$$
static pressure range: 0.5 to 8 inches water, column (in. w.c.)
material of construction: carbon steel
reference (= base date): **July 1988**

Each cost includes the fan, belt-driven motor, and motor starter.

For a given wheel diameter, the costs provided by Equation 4.7 will vary by up to ±10%, depending on the size of the motor sold with the unit. However, this variance falls within the accuracy range of the costs— ±20%.

Two vendors also supplied current prices for backward-curved blade fans *without* motors. These fans are fabricated of FRP (fiber-reinforced plastic).[5,6]

These prices include the fan (Class II construction) with a hand-layup, solid FRP wheel, all mounted on a coated steel base. The cost correlation is:

$$\text{Price (\$)} = 53.7D^{1.38} \tag{4.8}$$

where
$$10.5 \leq D, \text{ in. } \leq 73$$
reference (= base date) **April 1988**

Radial (Straight)–Blade Centrifugal Fans

Finally, we obtained vendor quotes (reference date: **July 1988**) for a range of radial-tip fan sizes.[7] These prices, which were based on welded, carbon steel construction and an operating temperature limit of 1000°F, were obtained for two groups of fans. Each group corresponded to a static pressure range. The price correlations were of the form:

$$\text{Price (\$)} = aD^b \tag{4.9}$$

where a and b are parameters.

The parameters, static pressures, and applicable ranges of these correlations are given below:

Group	"A"	"B"
Static pressure (in.):	2 to 22	20 to 32
Gas flowrate (acfm):	700 to 27,000	2000 to 27,000
Correlation parameters:		
a	6.41	22.1
b	1.81	1.55
Range in D (in.):	19.125 to 50.5	19.25 to 36.5

Example: A fan is needed to "push" 11,000 acfm through a system with a 10-in. pressure drop at standard conditions. The gas contains a high dust loading. Select a fan and estimate its price.

Solution: From Figure 4.1, we see that the familiar model 454 LS radial-tip fan can fill the bill. With a 45-1/8-in. wheel diameter, it falls into Group "A" above. Therefore:

$$\text{Price (\$)} = 6.41(45.125)^{1.81} = \$6330$$

Figure 4.1 also shows that either model 404 LS or model 504 LS could also handle this stream. We'll let you determine whether either (or both) is less expensive than model 454 LS.

Motors, Drives, and Starters

For non-packaged fans, the estimator must obtain prices for motors and other supporting equipment — starters, drives, inlet/outlet dampers, etc.

Damper prices will be covered in the "Ductwork" section. The following two equations provide prices for three-phase carbon steel motors for sizes ranging from 1 to 150 hp:[8]

$$\text{Motor cost} - \text{low hp (\$)} = 235(\text{hp})^{0.256} \qquad (4.10)$$

where $1 \leq \text{hp} \leq 7.5$

$$\text{Motor cost} - \text{high hp (\$)} = 94.7(\text{hp})^{0.821} \qquad (4.11)$$

where $7.5 \leq \text{hp} \leq 150$

Each price (reference date: **February 1988**) includes the motor (three-phase), V-belt drives, belt guard, and starter. Add these costs to those provided by Equations 4.8 and 4.9 to obtain cost estimates of fans and supporting equipment.

Example: In the preceding example, what size motor will the fan require, what will be its efficiency, and how much will it cost?

Solution: First, in Figure 4.1 we see that, to convey 11,070 cfm flow-rate at 10 in. static pressure, model 454 LS requires 28.2 BHP. If we substitute this in Equation 4.2 and solve for fan efficiency (n'), we obtain:

$$n' = 0.0001575(11,070)(10)/(28.2) = 0.62$$

Finally, we obtain the motor price from Equation 4.11:

$$\text{Price} = 94.7(28.2)^{0.821} = \$1,470$$

Lastly, because the fan and supporting equipment comprise only a part— and usually a *small* part—of a control system's total equipment cost, the installation factors and economic life of the *entire* control system apply to them. But in those (rare) cases when the fan price must be estimated separately (such as when it is replaced by a new fan before the control system is retired), the following information may be used in the fan replacement cost calculation:[9,10]

- Equipment lifetime: 12 to 30 years
- Installation cost: 25 to 100% of fan price

Clearly, both the installation cost and the equipment lifetime span broad ranges, as these parameters vary considerably with changing installation conditions and fan specifications, respectively.

DUCTWORK

Ductwork is needed to convey the waste gas stream from the emission source to, through, and from the control device. This section deals with ductwork selection, sizing, and costing.

Description

First, a definition: "Ductwork" includes: (1) straight duct (circular or rectangular), flanges, supports, fittings (elbow, tees, transitions, etc.), and dampers. Most or all of these items are used in a typical air pollution control system. Designing them for a given control system can be quite complex and time-consuming, and can require a great deal of source-specific information. (Reference 11 provides a thorough design procedure, for example.) When making a "budget-study" cost estimate, however, one does not always know the actual distance from the source to the control device—let alone the detailed ducting layout. Without knowing the layout, we can only guess at the numbers of fittings, supports, etc., required. Hence, for estimating purposes, it is adequate merely to base the ductwork cost on the *equivalent length of straight duct*. These duct lengths are typically "guesstimates" themselves (e.g., 100 ± 50 ft.). Clearly, given such imprecision in the input data, there is no need to "gild the lily" by laboriously determining the required size and number of fittings.

That said, what *kinds* of straight duct are used in control systems? Four kinds are most commonly used: (1) carbon steel, (2) stainless steel(s), (3) polyvinyl chloride (PVC), and (4) fiber-reinforced plastic (FRP). Steel ducting (either sheet or plate) is almost always custom-fabricated and, for larger sizes, must be built onsite. Although it is much heavier than either PVC or FRP duct and typically more expensive to fabricate, steel duct is fire-resistant and can accommodate quite high gas temperatures. Specifically, carbon steel duct can handle gases up to 1150°F; stainless steel, up to 1500°F. Carbon steel duct cannot handle corrosive gases, though stainless steels can at lower temperatures.[12]

The best selling points of PVC and FRP duct are their light weight and corrosion-resistant properties. These synthetics are either extruded (in diameters up to 2 ft.) or molded. While FRP duct fittings are straightforward to mold, certain PVC fittings are not so easy to fabricate. Lastly, the fire-resistant properties of synthetic duct are not laudable: when exposed to a flame, both PVC and FRP duct will melt. Worse yet, the PVC duct may decompose, emitting vinyl chloride and other toxic fumes. For this reason, most fire codes prohibit the use of PVC duct in exhaust systems conveying combustibles.

Sizing Procedure

As Table 3.1 indicates, the primary sizing parameters for ductwork are *length (L)* and *diameter (D)* OR *weight (W)*. To repeat, for purposes of making "budget-study" estimates, only a rough estimate of the total duct length is usually possible, as the precise length will depend on the source configuration. However, it is unusual for duct lengths to exceed 300 feet, while 100 to 200 feet is more representative. (Recall that this is the *total equivalent* duct length, including all fittings.)

For round duct, the diameter depends on two independent variables: the waste gas volumetric flowrate (Q, acfm) and the duct velocity (v, ft/min). Therefore, the duct diameter (D, in.) is:

$$D = 13.54(Q/v)^{0.500} \qquad (4.12)$$

The flowrate, Q, is measured at the source. But what about the velocity, v? The duct velocity is dictated both by the properties of the waste conveyed and cost considerations. (That is, the lower initial cost of small-diameter duct is offset by higher fan horsepower costs due to higher pressure losses.) The following guidelines may help:[13]

1. For light-density dust (gases, smoke; zinc, aluminum oxide fumes; flour, lint), recommended duct velocity is 2000 ft/min.

2. For light-to-medium–density dust (grain, sawdust; plastic, rubber dusts), recommended duct velocity is 3000 ft/min.

3. For medium-to-heavy–density dust (steel furnace, cement, sand blasting, grinding, etc.), recommended duct velocity is 4000 ft/min.

4. For heavy-density dust (metal turnings, lead and foundry shakeout dusts), recommended duct velocity is 5000 ft/min.

Generally, knowing the diameter and velocity is enough to price the ductwork. But in other cases, duct is priced on a dollars-per-pound basis, so that the weight must also be determined. For 1/4-in. wall thickness duct, calculate the *base* duct weight (W, lb/ft) from this expression:

$$W = 32.0dL \qquad (4.13)$$

where   d = duct diameter (ft)
        L = duct length (ft)

This equation gives only the weight of carbon steel plate in the duct. Additional steel must be added to strengthen and support the duct wall ("stiffeners"). A vendor (who wishes to remain anonymous) provided these multipliers for estimating the total duct weight:

Static Pressure (in. w.c.)	Stiffener Multiplier
0–10	1.15
10–20	1.225
20–30	1.30
30–40	1.35

Example: A furniture plant sanding machine exhausts 15,400 acfm of sawdust-laden air to a cyclone located 150 feet away. What are the diameter and weight of the (1/4-in. duct) required? Assume a maximum static pressure of 10 inches.

Solution: The above velocity table suggests a 3000 ft/min duct velocity. Substituting this in Equation 4.12, we get:

$$d = \text{diameter} = 13.54(15,400/3,000)^{0.5} = 2.56 \text{ ft}$$

From Equation 4.13:

$$\text{Base duct weight} = (32.0)(2.56)(150) = 12,300 \text{ lb}$$

And for a 10-in. w.c. maximum static pressure:

$$\text{Total duct weight} = 1.15 \times 12,300 \text{ lb.} = 14,100 \text{ lb}$$

Estimating the duct initial price is only part of the estimator's task. He also has to estimate the duct *pressure drop*, for purposes of determining its contribution to the control system power consumption. Equation 4.14 predicts this pressure loss:

$$\Delta P_d = 1.38 \times 10^{-7}(Q^{-0.5})(v^{2.5}) \tag{4.14}$$

where $\quad \Delta P_d$ = duct static pressure loss (in./100 ft).

This equation, derived from data in reference 14, applies to duct with "moderate" inner wall roughness. Notice that Equation 4.14 permits calculation of pressure loss with the knowledge of only *two* parameters: velocity and volumetric flowrate.

Example: Given that the above furniture plant cyclone has a pressure drop of 8 in. w.g., is the assumed maximum duct static pressure of 10 in. reasonable? Assume that the rest of the pressure losses total 1.0 in.

Solution: If it were reasonable, the static pressure loss in the ductwork would have to be 1 in. or less (i.e., 10 - 8 - 1 = 1). Now, Equation 4.14 tells us that:

$$\Delta P_d = 1.38 \times 10^{-7} (15,400)^{-0.5} (3,000)^{2.5} = 0.55$$

Ductwork losses = 0.55 in./100 ft × 1.5 = 0.83 in.

Therefore, our maximum static pressure assumption is valid.

Like straight duct, dampers are sized by their diameter or the diameter of the duct in which they are installed. Dampers are mainly used to regulate gas flow within duct systems. Fans also use inlet and outlet dampers to regulate the volumetric flowrate passing through them.

Costing Procedure

Vendors have supplied prices for PVC, FRP, carbon steel, and stainless steel fabricated duct. The correlations we developed from these prices (all in **August 1988** dollars) are shown below:

PVC Duct

Two correlations were developed, each to fit a different diameter range. Each equation is of the form:

$$(Cost\ (\$/ft.)\ =\ aD^b \tag{4.15}$$

The parameters for Equation 4.15 are:

Diameter range (in.):	6–12	14–24
Correlation parameters:		
a	0.877	0.0745
b	1.05	1.98

(*Note:* These prices were supplied by a vendor who did not wish to be identified.)

FRP Duct[15]

The price correlation here is the simplest:

$$Cost\ (\$/ft)\ =\ 24d \tag{4.16}$$

where $2 \leq d\ ft \leq 5$

Notice two features about these equations. In Equation 4.15, the diameter (D) is expressed in *inches*, while the diameter (d) is in feet in Equation 4.16. Also, note that the diameter ranges in the equations do not overlap—i.e., 6 in. to 2 ft. and 2 to 5 ft. This is not fortuitous. In general, PVC is less costly to fabricate than FRP, *for diameters under 2 feet.* For larger diameters, FRP is less costly.

Carbon and Stainless Steel Duct

Unlike its FRP and PVC counterparts, steel plate duct is almost always custom-fabricated, so that price lists are rarely compiled. (However, as shown below, steel *sheet* duct is often pre-fabricated.) Still, using historical data we can develop budget prices for steel plate ductwork. A ductwork fabricator (who prefers anonymity) supplied such prices on a dollars-per-pound basis, for both carbon and stainless steel plate straight duct and fittings (elbows and reducers). These data were:

Duct Diameter (ft)	Price ($/lb)
2–4	1.03
>4	0.97

These prices pertain to ASTM A-36 steel plate round duct with 1/4-in. wall thickness. The prices of fittings (round) were given as multiples of the straight duct price. These multipliers were:

Fitting		Price
Elbows:	≥ 90	1.65
	< 90	1.55
Reducers:	concentric	1.32
	eccentric	1.10

Finally, to obtain *estimates* of prices for straight duct and fittings made of stainless steel, multiply the carbon steel price by these factors:

Stainless Type	Multiplier
304	3.6
304L	3.7
316	4.3
316L	4.4

Prices for steel *sheet* straight duct were obtained from a fabricator for three wall thicknesses: 16-, 14-, and 11-gauge, which correspond to 1/16-, 5/64-, and 1/8-in., respectively.[16] These prices (in **February 1989** dollars) were regressed against the duct diameter (inches) to fit equations of the form of Equation 4.15. These correlations applied to duct diameters ranging from 6 to 40 in. The correlation parameters for these equations were:

Wall Thickness (ga.)	Straight-Duct Parameters	
	a	b
16	1.67	0.771
14	1.88	0.790
11	2.10	0.839

This same fabricator also supplied prices for 90-degree elbows and three-way, 90-degree tees. The elbows were "7-piece" and had centerline radii of 1.5 times the duct diameters. The regressed prices also fit Equation 4.15 and applied to diameters ranging from 11 to 40 in. Correlation parameters were:

| | Elbow Parameters | |
Wall Thickness (ga.)	a	b
16	2.60	1.50
14	3.19	1.48
11	4.52	1.43

Lastly, the correlation parameters for 90-degree tees were:

$$a = 13.8 \text{ and } b = 0.751$$

This correlation is for 16-ga. tees and applies to diameters ranging from 3 to 30 in.

A final note: All of the above prices pertain to carbon steel sheet. To obtain prices for *galvanized* sheet, add 35% to any of the above costs.

Dampers[17]

We obtained costs for two kinds of dampers—"back flow" dampers and "two-way diverter valves." A *back flow* damper is simply a circular piece of carbon steel installed in a duct via a steel rod fastened to it. One end of the rod protrudes outside the duct, allowing one to control the gas flow by rotating the metal disk in the duct. Fully closed, the damper face sits perpendicular to the gas flow direction; fully open, the face is parallel to the flow lines.

The cost data supplied fit the following correlation:

$$\text{Damper Price (\$)} = 7.46(D^{0.944}) \tag{4.17}$$

where $10 \leq D, \text{in.} \leq 36$

Reference (= base) date is **February 1988**.

Unlike back flow dampers that are installed in a section of straight duct, *two-way diverter valves* are installed in "Y" fittings, where they are used to totally or partially divert the gas flow from one branch of the Y to the other. Like the back flow dampers, two-way diverter valves are controlled manually.

These cost data also fit a power function, viz.:

$$\text{Diverter cost (\$)} = 4.84(D^{1.50}) \tag{4.18}$$

where $13 \leq D, \text{in.} \leq 40$

STACKS

Stacks are installed after control devices to disperse waste gases above ground level and surrounding buildings. That is their primary use. In the past, some stacks—in particular, *tall* stacks—have been used in lieu of control devices to dilute outlet pollutant concentrations, so as to meet applicable ambient air quality regulations. However, such usage has been severely constrained by the "tall stack" regulations promulgated by the U.S. Environmental Protection Agency.

Description

For discussion purposes, stacks are divided into two categories: "small" (those shorter than 100 ft) and "tall" (100 ft and higher). The sizing and costing procedures for these differ so much that it's convenient to treat them separately.

Small stacks are typically fabricated of carbon steel or FRP. Depending on the waste gas temperature, carbon steel stacks may or may not be lined with refractory. These stacks are essentially vertical ducts erected on foundations and supported by steel cables (if free-standing) or fastened to adjacent structures. For structural stability, the diameter of the stack bottom is slightly larger than the stack top. (Because the stack exit velocity is measured at the top, the top diameter is used for sizing purposes.) Small stack design is influenced by such factors as:[18]

- desired exit velocity
- maximum wind loading
- seismic zoning
- soil-bearing characteristics
- building code requirements

The same factors influence the design of *tall stacks* ("chimneys")—only more so, as these stacks are built to heights up to 1000 feet and more. In fact, the design and construction of some tall stacks rivals that of an entire plant in terms of complexity, magnitude, and expense.

Tall stacks consist of an outer shell and a liner. The outer shell is usually formed from reinforced concrete, though it may also be from brick. Either *steel* or *acid-resistant brick* is used for the liner. Steel-lined chimneys are usually used when the flue gas is above the acid (H_2SO_3/H_2SO_4) dew point—typical of unscrubbed waste gases and exhaust from dry flue gas desulfurization (FGD) systems and fluidized bed boilers. Steel liners consist of carbon steel, except for the top section (1.5 stack diameters in length) which is 316L stainless steel. Acid-resistant brick liners are used with flue gas from wet FGDs and in other situations where the flue gas temperature drops below the acid dew point.[19]

Sizing Procedure

For budget-study estimating purposes, the parameters we have to specify are the *stack diameter* and *stack height*. As stated above, the stack diameter is the *inside* diameter, measured at the stack exit. This diameter may be calculated from Equation 4.12 above:

$$D \text{ (in.)} = 13.54(Q/v)^{0.500}$$

where Q = flue gas exit flowrate (acfm)
        v = flue gas exit velocity (ft/min)

To obtain adequate dispersion, the *minimum* exit velocity should be 1.5 times the expected wind velocity. (For example, with a 50 mi/hr wind, the minimum velocity should be 6600 ft/min.) However, the exit velocity should be no higher than 9000 ft/min.[20]

Estimating the stack height is not as easy as calculating the stack diameter. The height required will depend on such variables as the height of the source; the stack exit velocity and temperature; the height, shape, and arrangement of nearby structures and terrain; and the pollutant concentration(s) at the stack outlet. For sources subject to State Implementation Plans (SIP), the stack height must be selected according to "good engineering practice" (GEP). For stacks constructed after January 12, 1979, the GEP stack height shall be the *greater of*: (1) 65 meters (213 ft); (2) the height demonstrated by an approved fluid model or field study that ensures that stack emissions do not cause excessive pollutant concentrations from atmospheric downwash, wakes, eddy effects, etc.; or (3) the height determined by the following equation:[21]

$$H = H_s + 1.5L \tag{4.19}$$

where H = GEP stack height, measured from the ground level elevation at the stack base (ft)
        H_s = height of nearby structure(s) measured from this ground level elevation
        L = lesser dimension (height or projected width of nearby structure(s)

Because these guidelines have been based on extensive fluid modeling and field study, they also are suitable for sources not subject to SIPs.

Costing Procedure

Small Stacks—Steel

Stack costs are difficult to generalize, as stacks are invariably custom-built to suit site-specific conditions. Small stacks are essentially sections of straight duct erected vertically and supported. Therefore, as a first estimate, we can obtain the stack steel "shell" cost from the straight duct cost in the previous section and multiply this cost by a factor to account for the cables, flanges, clamps, and other required accessories. The product of the duct cost and this factor would be the stack cost. We developed such multipliers from vintage duct and stack cost data.[22,23] We found that, overall, the stack/duct cost ratio ranged from 1.0 to 2.3. The ratio decreased as both the stack (duct) diameter and length *increased*. For example, for 50-ft tall stacks, the multiplier averaged 1.3, over a 2- to 5-ft diameter range. In fact, this multiplier (1.3) can be used as a first approximation, for steel stacks between 20 and 90 ft in height and 2 and 5 ft in diameter.

Small Stacks—FRP

The same costing procedure applies here. However, one vendor did indicate that FRP stack costs are approximately \$3/in.-ft, in **August 1988** dollars.[24] And, as Equation 4.16 indicates, the price of FRP duct ($\geq$ 2-ft diameter) is \$2/in.-ft. Therefore, the equivalent stack/duct cost ratio (multiplier) for FRP stacks would be 1.5 (i.e., \$3/\$2). This compares well with the 1.3 multiplier given above.

Example: The waste gas stream from a cement kiln is controlled by a fabric filter, then exhausted to a stack located at ground level. The stream flowrate is 25,000 acfm at 300°F. The maximum wind speed in the area is 60 mph. The tallest nearby building is 30 ft tall, while the "projected width" of adjacent buildings is 40 ft. Either a carbon steel or a FRP stack may be used. Which would be less expensive? (Assume a 1/4.-in. stack wall thickness.)

Solution: At a maximum wind speed of 60 mph, the minimum stack exit velocity would be: 1.5×60 mph = 7920 fpm. Substituting this velocity in Equation 4.12, we get a (top) stack diameter of 24.1 in. — say, 2 ft. Next, we can estimate the stack height (H) from Equation 4.19:

$$H = 30 + (1.5)40 = 90 \text{ ft}$$

Stack costs (uninstalled) would be:

$$\text{FRP: } \$3/\text{in.-ft} \times 90 \text{ ft} \times 24 \text{ in.} = \$6480$$

$$\text{Carbon steel: } 1.3 \times \text{straight duct price}$$

And: duct price = \$1.03/lb × duct weight.
From Equation 4.13:

$$\text{Weight (lb)} = 32.0 \times 2 \text{ ft} \times 90 \text{ ft} = 5760 \text{ lb}$$

(*Note:* no stiffening factor has been included in the duct weight calculation because the stack/duct price adjustment factor already accounts for it.)
Thus:

$$\text{Steel stack cost} = \$1.03/\text{lb} \times 1.3 \times 5760 \text{ lb} = \$7710$$

Table 4.1   Tall Stack Cost Correlation Parameters[a]

Lining Type	Diameter (ft)	Parameters a	b
Steel[b]	15	0.0120	0.811
	20	0.0108	0.851
	30	0.0114	0.882
	40	0.0137	0.885
Brick[c]	15	0.00602	0.952
	20	0.00562	0.984
	30	0.00551	1.027
	40	0.00633	1.036

Caution: Do NOT extrapolate parameters outside height range of correlations (200 to 600 ft).

Source: Zurn Constructors, Inc. All costs in **March 1988** dollars.
[a]Parameters pertain to cost equations of the form:
    Cost (million $) = aH^b
        where H = stack height (ft)
[b]Carbon steel liner with 316 L stainless steel top section.
[c]Acid-resistant firebrick liner.

As these costs show, the FRP stack is somewhat cheaper than the carbon steel stack.

Tall Stacks

Where the cost of small stacks is relatively modest, with tall stacks (pardon the expression), "the sky is the limit." Investments for tall stacks usually start at $1 million, with costs of $4 to $5 million not being uncommon.

Based on data taken from actual projects, Zurn Constructors supplied costs for tall stacks with both steel and acid-resistant liners, covering a range of diameters and heights. Unlike the above small stack costs, which include equipment only, the tall stack costs are *total capital investment* figures. In **March 1988** dollars, they include all labor and material costs associated with erecting the stack onsite, including such standard accessories as aircraft warning lights, ladders, and platforms.[25]

We regressed these costs against the stack height for each of four diameters (15, 20, 30, and 40 ft) and for either steel or brick lining. Each correlation is of the form:

$$\text{Cost (million \$)} = aH^b \qquad (4.20)$$

where   H = stack height (ft)
            a, b = correlation parameters

These correlation parameters are listed in Table 4.1.

Example: Estimate the capital cost of a 25-ft–diameter, 550-ft steel-lined stack. This diameter falls between the diameters for which we have correlations. Thus, we'll have to interpolate:

At 20-ft dia.: TCI = $0.0108(550)^{0.851}$ = $2.32 million

At 30-ft dia.: TCI $= 0.0114(550)^{0.882} = \$2.98$ million

Since 25 ft is exactly halfway, we can interpolate by simply averaging these two costs, or:

TCI (at 25-ft dia.) $= \$2.65$ million

CYCLONES

Cyclones (also known as "mechanical collectors") may be used as primary control devices, but more commonly they serve as precleaners to remove the bulk of heavier particulate from a waste gas stream before it enters the main control device.

Description

Cyclones are classified according to how the inlet gas enters the unit — i.e., *tangentially* or *axially*. In tangential entry cyclones, the waste gas enters an opening located on the tangent of the top of the unit. This causes the gas to whirl in a vortex through the upper body of the cyclone. Meanwhile, the particles, thrown against the inner wall by centrifugal force, settle to the bottom and fall into a dust hopper there. In axial flow cyclones, the gas enters at the middle of one end of a cylinder and flows through vanes that cause the gas to spin. A peripheral stream removes collected particles, while the cleaned gas exits at the center of the opposite end of the cylinder. Of the two designs, tangential entry cyclones are those most commonly used in air pollution control.[26]

As cyclones contain no moving parts, they are simple to operate and relatively maintenance-free. However, for optimum operation, the interior surfaces should be kept smooth and clean. Figure 4.2 depicts a tangential entry cyclone.

Cyclones may be built and used singly, in parallel, or in series. A large number of very small cyclones used in parallel is called a "multiclone" which, if properly designed, can achieve relatively high removal efficiencies — up to 80% of 5-μm particles.[27] Consequently, multiclones have been used as *primary* control devices in many applications.

Sizing Procedure

Cyclone particulate removal efficiency is a function of the particle density and diameter distribution, the cyclone diameter and geometric configuration, the gas volumetric flowrate, and other variables. But in general, most cyclones remove virtually every particle in the 20- to 30-μm range.[28] To characterize cyclone operation, it is convenient to define the *critical particle size* — the size of the *smallest* particle the cyclone can remove at 100% efficiency. Knowing the inlet particle size distribution, one can arbitrarily select a critical particle

Figure 4.2. Cyclone with dust collection drum (courtesy Sternvent Co., Inc.).

size and then determine the dimensions of a cyclone that can remove 100% of all particles that size and larger.

The key dimension in (and primary sizing parameter of) a cyclone is the *inlet area* (A, ft^2), which we can calculate from the following equation:[29]

$$A = [Q(r_p - r)/u]^{1.33}(d_c^{2.67}) \qquad (4.21)$$

where d_c = critical particle size (μm)
         u = gas viscosity (lb/ft-sec)
         r_p, r = densities of particles and gas, respectively (lb/ft^3)
         Q = gas flowrate (acfm)

However:

$$A = Q/v_i \qquad (4.22)$$

where v_i = the cyclone inlet velocity (ft/min).

Typically, v_i ranges from 4000 to 6000 ft/min. Thus, by selecting an inlet velocity, one can use Equation 4.21 to estimate the critical particle size for a given volumetric flowrate, viscosity, etc.

The inlet velocity also determines the cyclone pressure drop (P, in. w.c.):[30]

$$P = 2.36 \times 10^{-7}(v_i^2) \qquad (4.23)$$

But, Equations 4.21 and 4.23 hold only for a standard cyclone geometry and a "typical" number of gas stream revolutions in the unit.

Costing Procedure

A vendor provided cyclone costs for a range of inlet areas.[31] Each cost includes a carbon steel cyclone, support stand, and a fan and motor for pulling the waste gas through the unit. Also included is a rotary air lock and either a drum or a hopper for collecting captured dust, the drum being used for gas flows less than about 1000 acfm. These costs yielded the following correlations as a function of the inlet area, A:

1. Cyclone, fan, motor, supports, hopper/drum:

$$\text{Price (\$)} = 6520(A^{0.903}) \qquad (4.24)$$

where $0.200 \leq A, \text{ft}^2 \leq 2.64$

2. Rotary air lock only:

$$\text{Price (\$)} = 2{,}730(A^{0.0965}) \qquad (4.25)$$

where $0.350 \leq A \leq 2.64$

The costs in these equations are in **August 1988** dollars, their reference (and base) date. The *sum* of the costs yielded by Equations 4.24 and 4.25 is the price for the complete cyclone unit. Notice that the rotary air lock price correlation extends down to only 0.350 ft². That is because the smallest units, which collect dust in removal drums, do not require air locks. (See next section for additional rotary air lock prices.)

Example: Remember the furniture plant cyclone? What would the equipment in this control system cost – including the ductwork?

Solution:

1. *Cyclone*: Remember that the cyclone pressure drop was 8 in. w.c.

and the inlet flowrate was 15,400 acfm. Equation 4.23 tells us that this corresponds to an inlet velocity of:

$$v_i = (8/2.36 \times 10^{-7})^{0.500} = 5820 \text{ ft/min}$$

Thus: Inlet area (A) = 15,400/5820 = 2.65 ft^2.

Upon substituting this area into Equations 4.24 and 4.25, we obtain the following prices:

Cyclone, fan, motor, etc.: $15,700
Rotary air lock: 3,000
Total cyclone unit: $18,700

2. *Ductwork*: The total duct weight calculated above is 14,100 lb. Assume that 1/4-in. carbon steel duct is used. Because the duct diameter (31 in.) falls between 2 and 4 ft, the unit price to use is $1.03/lb or:

Duct price ($) = $1.03/lb $\times$ 14,100 lb = $14,500

The total control system equipment cost (in **August 1988** dollars) is the sum of these costs, or: $33,200.

ROTARY AIR LOCKS AND SCREW CONVEYORS

These two seemingly diverse devices work together to insure that dust captured by cyclones, fabric filters, and other dry collectors is efficiently removed and transported.

Description and Sizing Procedure

Rotary air locks are motor-driven dampers installed at the bottom of dust collection hoppers. They are designed so as to permit the dust to be withdrawn (intermittently or continuously) without allowing air to be pulled into the hopper and collector. Air locks are sized according to their diameter, which usually matches the diameter of the hopper outlet. Further, because the dust is usually non-corrosive, carbon steel fabrication is the norm.

Dust can be removed from hoppers manually if the inlet dust loading is low to moderate. In such cases, the dust would be emptied into drums or dumpsters. However, for higher dust loadings (say, 1 grain/dry standard ft^3 or more), continuous collection via screw conveyors is recommended. *Screw conveyors* consist of a screw fitted in a covered "trough" and linked to a motor via a drive. The captured dust flows through the air lock onto the continuously turning screw which moves (conveys) it along the trough, either horizontally, on an incline, or vertically. At the end of the trough, the dust is unloaded through a "discharge spout" — to a storage pile, truck, silo, or other endpoint.[32] (Figure 4.3 illustrates a typical screw conveyor.)

Screw conveyors are sized according to the *screw diameter*, the *conveyor length*, and (to a lesser extent) the *drive horsepower*. The diameter required

Figure 4.3. Screw conveyor system (courtesy Screw Conveyor Corp.).

depends on the type and quantity of dust conveyed (ft^3/hr), as well as the screw rotational speed (rpm) and other factors, such as the abrasiveness of the material. Screw diameters typically range from 6 to 24 in. The conveyor length depends on the distance between the hopper and the dust endpoint. Although conveyors can be built to any length, generally they are limited to approximately 150 ft. Lastly, the drive horsepower requirement depends both on the conveyor length and the volume loading.[33]

Costing Procedure

Rotary Air Locks

Using vendor-supplied data, we correlated rotary air lock equipment prices with their diameter (D, in.):[34]

$$\text{Price (\$)} = 429(D^{0.672}) \tag{4.26}$$

where $8 \leq D \leq 20$

Reference (and base) date: **February 1988**.

The above price includes the air lock, motor, chain drive with sprocket, sprocket guard, mount, and flanges on each end. *No* motor starter is included, however. (*Note:* Do not confuse these costs with the air lock costs presented above. The latter were costed specifically for the cyclones on which they were installed. Conversely, the air lock costs given by Equation 4.26 are generic.)

Screw Conveyors

Cost data supplied by Screw Conveyor Corporation[35] indicate that the total cost of a screw conveyor is the sum of the costs of the conveyor and the drive. The first and last terms in the following equation provide the costs of the conveyor and drive, respectively:

$$\text{Price} = 40.9L^{0.960} + 798(\text{hp})^{0.234} \qquad (4.27)$$

where L = conveyor length, ft ($50 \leq L \leq 150$)
         hp = drive horsepower ($1 \leq \text{hp} \leq 5$)

The cost of the conveyor in this equation includes: 6-in. diameter screws, tail shaft, trough ends, coupling shafts, discharge spout, supports, and fasteners. Meanwhile, the drive cost includes: 50- rpm TEFC motor, Class II screw conveyor reducer, V-belts, and belt guard. The drive horsepower requirement depends on the variables noted above. For instance, for a 60–70 lb/ft³ dust loading the screw at 1 and 5% of its capacity, the motor size required is 1 hp per 50 ft of conveyor length, up to 150 ft. However, when the loading increases to 25% of capacity, the drive horsepower approximately doubles — to 2 and 5 hp for 50- and 150-ft conveyors, respectively. (For more information on determining the drive horsepower, consult reference 32.)

HOODS

There is a truism in the air pollution control field: "No control device can remove pollutants that never get to it." Though this goes without saying, we often pay a lot of attention to the design and operation of a control device, while giving short shrift to the means for *capturing* the emissions at the source. In fact, the capture device is often the weakest link in the control system chain.

Capture methods fall into two categories: (1) *direct exhaust* and (2) *hoods*. In direct exhaust capture, the emission source is directly connected to the system ductwork via a duct fitting — usually an elbow. Thus, no special design is required. Costs for direct exhaust devices are given in the "Ductwork" section of this chapter. With hoods, however, matters are not as facile.

Description

Most capture hoods are either (1) *canopy* or (2) *semiclosed. Canopy hoods*, the more commonly used of the two, are mounted at some distance from the emission source. Consequently, they collect large amounts of ambient air along with the pollutants. Canopy hoods can be rectangular or circular, low or high. By convention, a "low" hood is one whose opening is no farther than three feet from the source – or a distance equal to the source diameter, whichever is less. All other canopy hoods fall into the "high" category.[36]

Semiclosed hoods avoid this problem, as they are normally located close to the source. However, some ambient air still does leak in through openings in the hood enclosure. The farther the hood is from the source, the higher the air infiltration. And because the costs of a control system – both capital and annual – are proportional to the *total* gas volumetric flowrate, it behooves designers to locate the hood as close to the source as possible.[37]

Sizing Procedure

Canopy Hoods

The sizing procedure for canopy hoods varies according to whether the hooded source (process, vessel, etc.) is hot or cold. The procedural differences are so great that they must be considered separately.

Cold process hooding. The primary sizing parameter for cold process hoods is the *capture velocity* – the nominal velocity at which the pollutants (and induced air) should enter the hood to be efficiently captured. This velocity is usually induced via the control system fan, which literally pulls the gases into the hood. The capture velocity varies with the nature of the process and the conditions of the air surrounding it. The semi-quantitative guidelines given below are helpful.[38]

Emission Velocity from Source	Condition of Surroundings	Capture Velocity (ft/min)
Zero	Still	50–100
Low	Moderately still	100–200
Moderate	Rapidly moving	200–500
High	Very rapidly moving	500–2000

For efficient capture, cold process canopy hoods are sized so that they overlap the source by 40%. (Figure 4.4 illustrates this.) Hence, emissions that would otherwise escape around the edge of the hood are caught. This 40% safety factor is implicit in the required ventilation rate equation:[39]

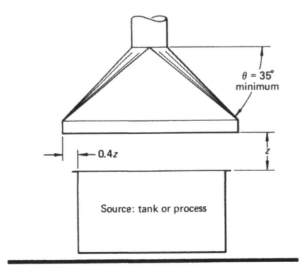

Figure 4.4   Cold-process low canopy hood. (Reprinted with permission from *Chemical Engineering*, December 1, 1980, pp. 111–115.)

$$Q = 1.4pzv \qquad (4.28)$$

where   Q = required ventilation rate to capture emissions (acfm)
          p = perimeter of hooded source (ft)
          z = open distance between source and hood (ft)
          v = capture (or "face") velocity (ft/min)

Therefore, Q is the flowrate that passes through the control system and is, in turn, the basis for the control system sizing and cost. (*Note:* if the cross drafts in the vicinity of the hood are minimal, no hood overlap would be needed, and the "1.4" factor in Equation 4.28 becomes simply "1.")

Hot process hooding. The design of hoods for hot processes is less cut-and-dried than sizing cold process hoods. Here, it is convenient to treat high- and low-canopy hoods separately.

Low canopies: Estimate the ventilation rates for circular and rectangular low-canopy hoods from the following two equations:

$$Q \text{ (circular-low)} = 4.7(d_h^{2.33})(\Delta T^{0.42}) \qquad (4.29)$$

$$Q \text{ (rectangular-low)} = 6.2L(w^{1.33})(\Delta T^{0.42}) \qquad (4.30)$$

where d_h = circular hood diameter (ft)
          ΔT = temperature difference between hot source and ambient air (°F)

L, w = length, width of rectangular hood, respectively (ft)

High canopies: The sizing of high-canopy hoods is complicated by such factors as the density difference between the source fumes and the ambient air. Moreover, as the hot gases rise toward the hood opening, they entrain air that acts to expand and cool the gases. The total flowrate of the "plume" (mixture of hot gases and ambient air) may be estimated as follows:

$$Q \text{ (circular-high)} = 7.4(z + 2d_s)^{1.5}(E^{0.33})) \qquad (4.31)$$

$$Q \text{ (rectangular-high)} = 18.5(z + 2w')^{0.59} \times \qquad (4.32)$$
$$[(l-w') + 0.5(z + 2w')^{0.88}](E^{0.33})$$

where d_s = diameter of source (ft)
   l, w′ = length, width of source, respectively (ft)
   E = heat transfer rate from source to plume (BTU/min)

Equations 4.29 to 4.32 were obtained from reference 40. Unlike Equation 4.28, which incorporates only spatial variables, these four equations also include thermal parameters, either a temperature difference or a heat transfer rate. The former can be measured rather directly. However, to determine E, the heat transfer rate, we need to know the temperatures of the plume and ambient air, the amount of air infiltration, degree of turbulence in the plume, and other hard-to-measure quantities.

Semiclosed Hoods

As complex as canopy hood sizing may seem, *semiclosed hood* design is ever more so. Semiclosed hoods are designed in various configurations to enclose the emission source. Hence, it is difficult to generalize their sizing and costing. Those interested in doing so should consult references 41 and 42.

Example: A circular vat of room-temperature sulfuric acid must be vented to a mist eliminator. The vat perimeter is 31.4 ft (i.e., 10-ft diameter). There are some weak cross currents in the immediate vicinity. A circular canopy hood is to be placed 2 ft above the vat. What must the hood diameter and ventilation rate be?

Solution: Because of the "mild currents" in the surrounding air, a 100- to 200-ft/min capture velocity will be needed, and Equation 4.28 will apply. Selecting the midpoint velocity, we have: v = 150 ft/min. As Figure 4.4 shows, the diameter of the hood must be the source diameter plus 40% of the vat-to-hood distance (2 ft), or:

$$d_h = 0.4z + d_s = 10.8 \text{ ft}$$

Finally:

$$Q = (1.4)(31.4)(2)(150) = 13,200 \text{ acfm}$$

Costing Procedure

The hood dimensions — diameter or length and width — are also the variables against which canopy hood costs are generally correlated. For instance, in the costing procedure presented in reference 43, the cost of both the hood sheet metal and the fabrication labor are correlated against the hood dimensions. The costs of the metal and fabrication labor are then summed to obtain the total hood cost. We can simplify this rather cumbersome procedure to get a single cost equation for each hood type, instead of two. That is:

Circular hoods:

$$\text{Price (\$)} = 47.1(d_h^{1.62}) \qquad (4.33)$$

where d_h = hood diameter (ft)
such that $3 \leq d_h \leq 60$

This equation applies only to circular hoods with 35-degree slopes and metal thicknesses of $\leq 3/16$ in. Hoods of the same diameter but with steeper slope angles would cost more.

Rectangular hoods:

$$\text{Price (\$)} = 39.0(L^{1.80}) \qquad (4.34)$$

where   L = hood length (ft)
such that $5 \leq L \leq 60$

Like Equation 4.33, this correlation pertains only to 35-degree sloped hoods with 3/16-in. or less thick carbon steel. Further, the applicable length/width (L/w) ratio is 1. For higher L/w ratios at any given L, the cost would be *lower*.

Finally, these costs have been escalated to an **August 1988** reference date from a December 1977 base, using two Bureau of Labor Statistics indices.[44] We inflated the metal prices using the "Metal Products" (Code 10) component of the Producer Price Index. And because wage rates are often contractually tied to the Consumer Price Index, we used the CPI to escalate the fabrication labor costs. The escalation increases for metal and labor were 77% and 96%, respectively.

An astute reader may notice that this escalation blatantly violates the guideline (commandment?) set forth in Chapter 3 — namely, that we should never escalate costs over periods longer than *five* years. We've violated it in this case, because the pricing of hoods is relatively straightforward, as their cost is comprised of just metal and fabrication labor. And published indices are available to escalate both costs. (Besides, there's an exception to every rule!)

Example: If the hood in the previous example had a 35-degree slope and were fabricated of 3/16-in. carbon steel, how much would it cost?

Solution: Equation 4.33 applies. Substituting the hood diameter therein, we obtain:

$$\text{Hood cost (\$)} = 47.1(10.8^{1.62}) = \$2,220$$

BUILDINGS

Some control systems are installed outdoors, while others are situated indoors, within the plant process area. Whether a control system is enclosed, partly enclosed, or unenclosed depends on such factors as available space, the need to protect the equipment and (especially) the instrumentation from the elements, and the required proximity of the control equipment to the process (e.g., in cases where the captured pollutants would be recycled to the process).

Description

Two general categories of buildings are used to house control equipment: *portable* and *permanent*. Portable buildings, in turn, may be subdivided according to *pre-assembled* and *site-assembled*. As their name implies, pre-assembled buildings are built at the factory and shipped fully assembled, except for the roof, gutters, and some accessories (e.g., air conditioners) which require field installation. A typical "base" pre-assembled building includes a floor, sliding windows, electrical outlets, a circuit breaker box, doors, and inside shelving. The roof, insulation, heating, etc., are extra. Due to "wide load" highway regulations, the shortest dimension (e.g., width) of a pre-assembled building is limited to 8 ft. However, another dimension (e.g., length) may be as large as 16 ft. Pre-assembled buildings are designed for outdoor installation on a concrete slab foundation provided by the customer.

Site-assembled portable buildings are intended for *indoor* installations, such as a plant process area. Site-assembled units come in a variety of sizes, styles, and designs. For instance, one manufacturer sells several floorless units that each consist of wall panels, wiring studs, a steel door, a vinyl base, an acoustical ceiling, a steel dust cover, and joinery. What distinguishes these units from each other are differences in the thicknesses and materials used in fabricating the wall panels and in the strength of the studding used in supports. Because these units are modular, they can be built to enclose any indoor floor area. They also can be assembled in multi-story modules. Installation time and skill required are minimal. A portable unit can be used to enclose an entire control system, particularly if the system is small and compact. More likely, it would be used to enclose the control panel and other instrumentation, especially if the immediate environment were dusty or hostile.

Permanent buildings also vary in design, style, and price. A typical plant building is rather spartan, however, consisting of a concrete floor, walls, roof, access doors, and lighting. The building cost varies according to the type of frame (steel or bearing walls) or wall used. Wall types may consist of: concrete block, brick-on-concrete block, galvanized steel siding, or metal sandwich panels. Also, such amenities as basements, fencing, emergency lighting, paving, and dock levelers may be added at additional cost.

Sizing Procedure

Permanent buildings are traditionally sized on a floor area ("square footage") basis. For portable buildings, either floor area or wall length (in linear feet) is the sizing parameter. In both cases, however, certain building appurtenances (such as air conditioners) are *selected*, not sized. That is, their cost is usually independent of square footage or any other measure.

To determine the size of the building required to enclose all or part of a control system, first consider the dimensions of each piece of equipment — length, width, and height — and how these equipment items are to be arranged. Also, provide some space around the equipment to allow access by operating, maintenance, and other personnel. (A 6-ft perimeter is usually adequate, for example.) The total of the equipment and access space determines the total space requirement which, in turn, determines the dimensions of the enclosure required.

But before sizing the enclosure, we first have to determine if we really need one. In other words, is enough room available inside a plant building to allow a space allocation to the control system? Generally, space allocation is less expensive than an enclosure, as unit building costs ($/ft^2 or $/lin. ft) usually decrease with increasing floor area. Nonetheless, there may be cases where a dedicated structure is needed, such as a control system located at a remote distance from the emission source. Moreover, space allocation may not be possible where the control system is retrofitted to an existing source, for in-building space simply may not be available in those cases.

Costing Procedure

For budget-study estimating purposes, figure building base costs as the product of the unit price and the measure (square footage or perimeter length). Costs for extras can be either added in "a la carte" or figured as percentages of the base cost.

Portable Buildings—Site-assembled

The unit price of these is a function of the building design — specifically, the wall type and thickness and type of supporting structure. The prices (**August 1988** dollars) listed below are for *base equipment only*, F.O.B. the manufacturer:

Wall Panel Features	Approximate Cost ($/linear ft)
Vinyl hardboard (1.75-in.)	37
Vinyl hardboard (3-in.)	54
Steel-faced (3-in.), with heavy-duty steel frame	67

All of the above wall panels contain Kraft honeycomb cores. The cost of accessories varies according to the size of the building. For example, a typical electrical package costs from $200 to $1200. Lastly, these costs were provided by a portable building manufacturer who prefers to remain anonymous.

As stated above, building installation costs are minimal, though they would vary primarily according to the skill and wage rates of plant personnel or others charged with erecting the units. For estimating purposes, this cost shouldn't exceed 25% of the above base equipment prices.

Portable Buildings—Pre-assembled

The same anonymous vendor supplied costs for pre-assembled buildings. The base cost (F.O.B. manufacturer, **August 1988**) ranged from roughly $50 to $150/ft^2, for floor areas of 128 and 15 ft^2, respectively. The following correlation captures this relationship:

$$\text{Price (\$/ft)} = 556(A^{-0.518}) \tag{4.35}$$

where $15 \leq A \ (ft^2) \leq 128$

For instance, an 8 ft $\times$ 8 ft unit would cost $64.5/ft^2 $\times$ 64 ft^2 = $4100.

To this base amount we must add the cost of accessories. A typical accessories package would include a roof, painting, insulation, insulated glass, window screens, and heater(s). Depending on the building size, this package would add from 60 to 70% to the base price.

Because these units are pre-assembled, less installation would be required than for site-assembled buildings. However, as we said above, the customer must furnish the concrete pad on which the building is mounted. Therefore, to be conservative, we can use the same installation cost factor as above—i.e., 25% of the total equipment cost.

Finally, the life of both pre- and site-assembled portable buildings is relatively short, compared to that of permanent buildings. Recognizing this, the IRS allows one to depreciate portable buildings in just 7 years, compared to approximately 30 years for their permanent counterparts. This short life has to be weighed against the price, flexibility, and convenience advantages offered by portable buildings.

Permanent Buildings

Again, the unit price of permanent buildings depends on the wall and frame types, as well as the square footage. For instance, a building with a 30,000-ft^2 floor area and a concrete block bearing wall would have a unit capital cost of about \$38/ft^2.[45,46] We obtained ranges of unit costs for different exterior wall types (all with steel frames) from the same references:

Exterior Wall	Unit Capital Cost (\$/ft^2)
Brick-on-concrete block	41–59
Concrete block	36–47
Galvanized steel siding	38–48
Metal sandwich panels	36–48

For basements, *add* approximately \$13/ft^2 to the above costs. For buildings with "bearing wall" type frames, *deduct* \$1/ft^2 from the figures above.

The upper and lower costs in these ranges correspond to floor areas of 10,000 and 60,000 ft^2, respectively. Clearly, the unit cost depends as much (or more) on the floor area as on the wall or frame type. The reference date for these costs is **July 1988**; the base date, July 1985. These prices also would depend on the location of the building, due to geographic variations in construction wages.

*   *   *

So much for the auxiliaries. Although we've not covered all of the control device auxiliaries, we have included the most important ones. Others (e.g., pumps) almost always come with the control devices (e.g., wet scrubbers), so there is little need to present data for them. Still other auxiliaries, such as air compressors for pulse-jet fabric filters (see Chapter 5), are too specialized for coverage here. But what *does* appear in this chapter should be enough to satisfy the needs of most "budget-study" estimators.

REFERENCES

1. Danielson, J. A., Ed. *Air Pollution Engineering Manual*. Research Triangle Park, NC: U.S. Environmental Protection Agency, May 1973 (NTIS PB-225132) (hereinafter cited as *Engineering Manual*), pp. 60–62.
2. *Engineering Manual*, p. 67.
3. Calvert, S., and H. M. Englund, Eds. *Handbook of Air Pollution Technology*. New York: John Wiley & Sons, 1984 (hereinafter cited as *Handbook*), p. 350.
4. Price and technical data from L.R. Gorrell Company (Raleigh, NC), July 1988 (hereinafter cited as L.R. Gorrell).
5. Price and technical data from Hoffman and Hoffman (Raleigh, NC), April 1988 (hereinafter cited as Hoffman and Hoffman).
6. Price and technical data from Vanaire, Inc. (Louisville, KY), April 1988.
7. L.R. Gorrell.

8. Price and technical data from Murphy-Rodgers, Inc. (Huntington Park, CA), February 1988 (hereinafter cited as Murphy-Rodgers).

9. L.R. Gorrell.

10. Hoffman and Hoffman.

11. *Engineering Manual*, pp. 44–59.

12. Vatavuk, W. M., and R. B. Neveril. "Estimating Costs of Air-Pollution Control Systems, Part IV: Estimating the Size and Cost of Ductwork," *Chemical Engineering*, December 29, 1980, pp. 71–73 (hereinafter cited as "Ductwork").

13. "Ductwork."

14. *Engineering Manual*, pp. 45–46.

15. Price data from Skilcraft Fiberglass Corporation (Glendale, AZ), August 1988.

16. Price data from Wer-Coy Metal Fabrication Co. (Warren, MI), February 1989.

17. Murphy-Rodgers.

18. Vatavuk, W. M., and R. B. Neveril. "Estimating Costs of Air-Pollution Control Systems, Part VIII: Estimating Costs of Exhaust Stacks," *Chemical Engineering*, June 15, 1981, pp. 129–130 (hereinafter cited as "Stacks").

19. Price and technical data from Zurn Constructors, Inc. (Tampa, FL), June 1988 (hereinafter cited as Zurn Constructors).

20. "Stacks."

21. *Guideline for Determination of Good Engineering Practice: Stack Height (Technical Support Document for the Stack Height Regulations)* (Revised). Research Triangle Park, NC: U.S. Environmental Protection Agency, June 1985 (NTIS PB-85-225241), pp. 1–2.

22. "Ductwork."

23. "Stacks."

24. Price data from Air Plastics, Inc. (Cincinnati, OH), August 1988.

25. Zurn Constructors.

26. *Handbook*, p. 323.

27. Vatavuk, W. M., and R. B. Neveril. "Estimating Costs of Air-Pollution Control Systems, Part V: Estimating the Size and Cost of Gas Conditioners," *Chemical Engineering*, January 26, 1981, pp. 127–132.

28. *Handbook*, p. 348.

29. *Handbook*, p. 348.

30. *Handbook*, p. 348.

31. Price and technical data from Chet Adams Company (Cary, NC), August 1988.

32. *Screw Conveyor Catalog and Engineering Manual* (Catalog No. 787). Hammond, IN: Screw Conveyor Corporation, 1987.

33. *Screw Conveyor Corporation*.

34. Murphy-Rodgers.

35. Price data from Screw Conveyor Corporation (Hammond, IN), October 1988.

36. Vatavuk, W. M., and R. B. Neveril. "Estimating Costs of Air-Pollution Control Systems, Part III: Estimating the Size and Cost of Pollutant Capture Hoods," *Chemical Engineering*, December 1, 1980, pp. 111–115 (hereinafter cited as "Capture Hoods").

37. "Capture Hoods."

38. "Capture Hoods."

39. "Capture Hoods."

40. "Capture Hoods."

41. *Engineering Manual*, pp. 27–44.

42. *Industrial Ventilation* (latest edition available). Lansing, MI: American Conference of Governmental Industrial Hygienists.
43. *Handbook*, pp. 343–347.
44. Bureau of Labor Statistics, U.S. Department of Labor (Washington, DC), August 1988.
45. *Means Square Foot Costs 1985*. Kingston, MA: R.S. Means Company, Inc., 1985, pp. 208–209.
46. Price data from R.S. Means Company (Kingston, MA), August 1988.

CHAPTER 5

"Add-on" Controls II: Particulate Control Devices

> The dust comes secretly day
> After day
> Lies on my ledge and dulls my
> Shining things.
>
> Viola Meynell

In this and the next chapter we'll cover the sizing and costing of the most commonly used add-on air pollution control devices. Traditionally, engineers categorize add-ons according to the *state* of the pollutants they capture. *Particulate* control devices collect dust, mists, and other solid and liquid particles. *Gaseous* control devices, on the other hand, capture or convert gaseous pollutants. Some devices, such as wet scrubbers, can collect both particulate and gaseous emissions. But in general, the "division of labor" between the two categories is clear-cut. This chapter will deal with the "big three" particulate matter (PM) control devices: fabric filters, electrostatic precipitators (ESPs), and wet scrubbers, in that order.

Of these three, fabric filters and dry ESPs are "dry" collectors, so named because they capture the particulate in the dry state. Similarly, wet ESPs and wet scrubbers are termed "wet" collectors. This distinction used to be merely academic, but that was before the treatment and disposal of liquid and solid wastes became a major environmental concern. With the passage of RCRA (Resource Conservation and Recovery Act) regulations, wastes that once were summarily shipped to landfills, treatment ponds, and other disposal sites now have to be pretreated before disposal. And for certain wastes (e.g., polychlorinated biphenyls — "PCBs"), land disposal is *verboten*. For these outcasts, high-temperature incineration is the only permissible disposal method. Hence, the cost of treating and disposing of particulates often becomes nearly as large as the other annual costs of operating and maintaining the control system. For example, a recent publication lists waste disposal costs of $10 to $40/ton in municipal landfills and approximately *$250/ton* in RCRA landfills.[1] With such high costs abounding, it is no surprise that waste disposal costs can heavily influence the choice of the control device in some applications.

FABRIC FILTERS

Description

A wag once said, "A fabric filter is to dirty air what a vacuum cleaner is to a carpet." Arguably the most widely used PM control device, a fabric filter (or "baghouse") does just that: filter the particle-laden gas through fabric—specifically, cloth bags. The particles are caught on the surface of the bags, while the cleaned gas passes through to the ambient air. Typically, several bags are grouped together in a "compartment," while several such compartments comprise a fabric filter "unit." Filters are operated cyclically, with long periods of filtering alternating with short periods of cleaning. During cleaning, dust caught on the inside/outside of the bags (depending on the direction of gas flow) is removed, collected, and either recycled to the process, transported to storage, or disposed of.

Because the bag cloth is so finely woven (or felted)—and because the captured dust "caked" on the bags also acts as a filtering medium—fabric filters are quite efficient in removing particles much smaller than 1 micrometer or "μm" (10^{-6} meter) in diameter. In fact, baghouses typically remove 99+% of these "sub-micrometer" particles. Moreover, fabric filters can treat gas streams as hot as 500°F—provided that the moisture content and temperature are such that the gas remains above the dewpoint. Along with the moisture content, other process variables that affect baghouse design and operation are listed below.[2]

Particle Characteristics

The *adhesiveness* and *size distribution* of the incoming particles are the crucial parameters here. Adhering particles, such as oily residues or electrostatically active plastics, tend to "blind" the bag pores, thus inhibiting the gas flow and increasing the pressure drop. To combat this, a precoating material is injected onto the bag surface, which traps the adhering particles before they reach the bag surface. Smaller particles may not blind the bag pores, but they usually form a denser dust cake, which also increases the pressure drop and the fan power cost. We'll later see how particle size affects baghouse sizing.

Gas Stream Characteristics

Along with the moisture content, the amount of corrosive substances present also affects baghouse design. To deal with excessive moisture, the baghouse unit and ductwork can be either insulated to prevent heat loss or heated directly, as condensation adversely affects both the fabric and the baghouse structure. With corrosion, this effect is more pronounced—and more expensive to control. Typically, stainless steels are required for fabricating the baghouse compartment, ductwork, and other system components. (If appreciable chlorides are present, however, "austenitic" stainless steels will not suffice. More resistant materials would be needed.) Because the baghouse *operating temperature* is generally limited to 500°F, higher-temperature waste gas streams must be precooled before they reach the collector. Either a precooler (e.g., spray chamber) or dilution air can be used for this purpose. Adding dilution air increases the size (and costs) of the baghouse and auxiliary equipment, however.

Finally, the *operating pressure* of a baghouse is limited to ±25 inches water, column (in. w.c.). This limitation is due to the sheet metal construction used in the unit. For this reason, most baghouses are operated at (or near) atmospheric pressure.

Design Characteristics

This area can be subdivided according to the type of filter "housing" and type of construction. The housing may be either "pressure" or "suction." In pressure baghouses, the fan is located before (upstream) of the baghouse and conveys the dirty gas down into and out of the bags. Advantages offered include ease of spotting leaking bags and the use of a less expensive housing that requires no reinforcement for vacuum service. The down side is that the fan has to handle a dirty gas that is no friend to fan blades.

In "suction"-type housings, the fan sees only clean gas, as it is installed after the baghouse. But, the filter unit, to withstand relatively high negative pressures, has to be more heavily constructed and reinforced. Also, because the unit is under suction, outside air can be drawn in, causing condensation, corrosion, and even explosions, while increasing fan power and other operating costs. Finally, locating broken bags is not as easy in suction baghouses as it is in pressure units.

Both suction and pressure baghouses are built in both "standard" and "custom" units. Standard filters generally are off-the-shelf units that are vendor-assembled and shipped to the customer either with or without bags. The dust hopper and bag compartment sections are separate in larger standard units, but are "unitized" in the smaller baghouses. Very large (> 100,000 acfm) standard filters are often shipped in modules that are assembled onsite. But because they are pre-assembled, field installation labor is minimal.

"Custom" baghouses, which are also very large, are designed for specific applications. (See Figure 5.1.) Consequently, the unit cost (typically expressed in $/ft^2 of bag area) is much higher than the cost of standard filters. Custom baghouses have several advantages, however, such as ease of maintenance and easy access to bags needing replacement.

Filter Media Characteristics

The type of bag material needed depends on the gas chemical composition, temperature, and dust loading; the particulate physical and chemical charac-

Figure 5.1   Custom baghouse controlling a spreader stoker grate coal-fired boiler (courtesy Zurn Industries. Inc., Air Systems Division).

teristics; and, to some extent, the type of bag cleaning used. Bag fabrics may be *felted* or *woven* (sometimes knit) and are selected according to the yarn type (filament, spun, or staple), diameter, twist, and other factors. Generally, spun or heavier staple yarn fabrics are used with "shaker"-type cleaning mechanisms, while lighter-weight filament yarns are adequate for "reverse-air" cleaning baghouses. (Cleaning mechanisms are discussed below.)

The baghouse operating temperature has an especially big influence on the choice of fabric. At the low end (180°F), cotton fabric suffices, while fiberglass is needed at the high end (500°F). Table 5.1 lists the operating temperature, acid and alkali resistance, and "flex abrasion" properties of leading fabric materials. ("Flex abrasion" refers to the fabric's suitability to cleaning via mechanical shaking.) The Gore-Tex™ listed in Table 5.1 is not a fabric per se, but a membrane that is laminated to a substrate fabric to improve its filtering performance in extreme applications.

Cleaning Mechanism

Although the location of the system fan, direction of gas flow, and other factors help to characterize a baghouse, the bag cleaning method is the most notable. The three most commonly used cleaning methods are: (1) shaker, (2) reverse-air, and (3) pulse-jet.[3]

Used with inside-to-outside gas flow, *shaker cleaning* involves suspending each bag from a motor-driven hook or framework that oscillates during the cleaning cycle. These oscillations impart enough energy to the bag to overcome the forces holding the dust to the fabric. For smaller baghouses, the shaker

Table 5.1   Properties of Selected Bag Materials

Fabric	Temp (°F)[a]	Acid Resistance	Alkali Resistance	Flex Abrasion
Cotton	180	Poor	Very good	Very good
Creslan[b]	250	Good in mineral acids	Good in weak alkali	Good to very good
Dacron[c]	275	Good in most mineral acids; dissolves partially in concentrated H_2SO_4	Good in weak alkali; fair in strong alkali	Very good
Dynel[c]	160	Little effect even at high concentration	Little effect even at high concentration	Fair to good
Fiberglas[d]	500	Fair to good	Fair to good	Fair
Filtron[e]	270	Good to excellent	Good	Good to very good
Gore-Tex[f]	Depends on backing	Depends on backing	Depends on backing	Fair
Nomex[c]	375	Fair	Excellent at low temperature	Excellent
Nylon[c]	200	Fair	Excellent	Excellent
Orlon[c]	260	Good to excellent in mineral acids	Fair to good in weak alkali	Good
Polypropylene	200	Excellent	Excellent	Excellent
Teflon[c]	450	Inert except to fluorine	Inert except to trifluoride, chlorine, and molten alkaline metals	Fair
Wool	200	Very good	Poor	Fair to good

Source: Turner, J. H., A. S. Viner, R. E. Jenkins, and W. M. Vatavuk. "Sizing and Costing of Fabric Filters, Part I: Sizing Considerations," *Journal of the Air Pollution Control Association*, June 1987, pp. 749-759. (Reprinted with permission.)
[a]Maximum continuous operating temperatures recommended by the Industrial Gas Cleaning Institute
[b]American Cyanamid registered trademark.
[c]Registered trademark of E.I. duPont de Nemours and Co., Inc.
[d]Owens-Corning Fiberglas registered trademark.
[e]W.W. Criswell Div. of Wheelabrator-Fry, Inc., trademark.
[f]W.L. Gore and Co. registered trademark.

mechanism is controlled manually, while in larger, multicompartment filters, a timer or pressure sensor activates the cleaning apparatus. In both cases, the forward gas flow to the compartment is stopped, dust is allowed to settle, and the shaker mechanism is turned on for about a minute. This settling and shaking may be repeated, after which the compartment is brought back on-line for filtering. Because shaking is a vigorous operation, heavier and more durable bag fabrics (typically woven) are required.

Reverse-air cleaning is gentler than shaking. Here, gas flow to the compartment is stopped, dust is allowed to settle, and air is introduced to the compartment in a reverse direction via a separate fan. This air causes the bags to collapse and shed their dust via a shearing action. To prevent them from totally collapsing, rings are sewn into the bags.

Unlike their shaker or reverse-air counterparts, in *pulse-jet* baghouses the dusty gas flows from outside to inside the bags, which are mounted on wire cages to prevent their collapse. This allows them free movement during the

Table 5.2 Gross/Net Cloth Area Factors

Net Cloth Area (ft^2)	Gross Cloth Area (ft^2)—Multiply by:
1–4,000	2
4,001–12,000	1.50
12,001–24,000	1.25
24,001–36,000	1.17
36,001–48,000	1.125
48,001–60,000	1.11
60,001–72,000	1.10
72,001–84,000	1.09
84,001–96,000	1.08
96,001–108,000	1.07
108,001–132,000	1.06
132,001–180,000	1.05
above 180,001	1.04

Source: Turner, J.H., A. S. Viner, R. E. Jenkins, and W. M. Vatavuk. "Sizing and Costing of Fabric Filters, Part I: Sizing Considerations," *Journal of the Air Pollution Control Association*, June 1987, pp. 749-759. (Reprinted with permission.)

cleaning cycle, when a jolt of compressed air is shot into the top of the bags through quick-opening valves. The cleaning is done on-line, so that no section of the unit has to be shut down. (Typically, 10% of the baghouse is pulsed at any time.) Consequently, a pulse-jet baghouse requires less bag area than a shaker or reverse-air unit handling the same gas flow rate.

Sizing Procedure

The primary sizing parameter for a fabric filter is the *gas (or air)-to-cloth ratio*, typically expressed in acfm/ft^2 of filtering area, or "net cloth area." As alluded to above, in shaker and reverse-air baghouses, the net cloth area is less than the total ("gross") bag area. This is because extra bags must be installed to "take up the slack" while a compartment is being cleaned. Mathematically this is expressed as follows:

$$A(net) = Q/R \qquad (5.1)$$

where

$A(net)$ = net (filtering) bag area (ft^2)
Q = gas flowrate (acfm)
R = gas-to-cloth ratio (acfm/ft^2)

And:

$$A(gross) = A(net)f \qquad (5.2)$$

where f = factor to account for extra bag area needed

Values of "f" are listed in Table 5.2. Note that as the net bag area gets very large, f approaches unity. For pulse-jet units, which are cleaned on-line, and for "intermittent" baghouses, which are shut down completely for cleaning, f = 1.

How do we determine "R," the gas-to-cloth ratio? Answer: not easily, because R depends on many factors – particle and gas stream characteristics, fabric properties, type of cleaning, and others. Reference 4 contains a thorough discussion of these factors, especially the relationship between R and the pressure drop through the bags. This reference also discusses several methods for predicting R. We'll present two of them here.

In method #1, one selects a bag fabric, picks a gas-to-cloth ratio from Table 5.3, and substitutes this ratio into Equation 5.1. Note that Table 5.3 presents Rs for about 40 different dusts, from alumina to zinc oxide, for shaker/reverse-air and pulse-jet baghouses. Also note that no other particle, gas, or fabric characteristics enter into the selection. Nonetheless, these ratios are considered conservative and adequate for use in making "budget-study" cost estimates.

Method #2 is a "factor" technique developed by a baghouse manufacturer. It incorporates not only the dust type but also the nature of the emission source (e.g., screening), the dust fineness, and the dust loading. Table 5.4 encompasses these parameters and also provides (in the footnote) an example calculation. Note that this method is limited to *shaker* baghouses only.

Example: Emissions from a ball mill at a Portland cement plant are to be controlled by a shaker baghouse. The dust loading and mass median particle size range are 2.5 grains/scf and 1.7 μm, respectively. What gas-to-cloth ratio will we need?

Solution:

Method #1: From Table 5.3, we select a gas/cloth ratio of 2.0.

Method #2: Referring to Table 5.4, we first select a source code from Chart A of the table – "3," for "pulverizing." The "base" R would be 2.5/1 (column 3). From Chart B, select a particle fineness factor of 0.8, corresponding to a size range of 1 to 3 μm. Lastly, from Chart C, pick "1.2" for the 1–3 grains/scf dust loading range. The gas/cloth ratio, R, would be the product of these numbers, or: R = 2.5 × 0.8 × 1.2 = 2.4.

To be conservative, we should use the higher ratio (2.4).

Our final consideration is the baghouse pressure drop. Several semiempirical models have been developed to predict pressure drop as a function of the gas-to-cloth ratio, dust loading, filtering time, and such parameters as the "specific dust resistance" and the "effective residual drag" of the bag fabric. However, the models assume that one knows (or can accurately predict) these parameters. Without this knowledge, the equations are of little use. But, for estimating purposes, a *maximum* pressure drop of 5 to 10 in. w.c. can be used.[5]

Table 5.3   Gas-to-Cloth Ratios for Selected Dusts[a]

Dust	Shaker/woven, reverse-air/woven[b]	Pulse jet/felt[b]
Alumina	2.5	8
Asbestos	3.0	10
Bauxite	2.5	8
Carbon black	1.5	5
Coal	2.5	8
Cocoa, chocolate	2.8	12
Clay	2.5	9
Cement	2.0	8
Cosmetics	1.5	10
Enamel frit	2.5	9
Feeds, grain	3.5	14
Feldspar	2.2	9
Fertilizer	3.0	8
Flour	3.0	12
Fly ash	2.5	5
Graphite	2.0	5
Gypsum	2.0	10
Iron ore	3.0	11
Iron oxide	2.5	7
Iron sulfate	2.0	6
Lead oxide	2.0	6
Leather dust	3.5	12
Lime	2.5	10
Limestone	2.7	8
Mica	2.7	9
Paint pigments	2.5	7
Paper	3.5	10
Plastics	2.5	7
Quartz	2.8	9
Rock dust	3.0	9
Sand	2.5	10
Sawdust (wood)	3.5	12
Silica	2.5	7
Slate	3.5	12
Soap, detergents	2.0	5
Spices	2.7	10
Starch	3.0	8
Sugar	2.0	7
Talc	2.5	10
Tobacco	3.5	13
Zinc oxide	2.0	5

Source: Turner, J. H., A. S. Viner, R. E. Jenkins, and W. M. Vatavuk. "Sizing and Costing of Fabric Filters, Part I: Sizing Considerations," *Journal of the Air Pollution Control Association*, June 1987, pp. 749-759. (Reprinted with permission.)
[a]Generally safe design values—application requires consideration of grain loading and particle characteristics such as size distribution and electrostatic properties.
[b]Units are $(ft^3/min)/(ft^2$ of cloth area).

Costing Procedure

Equipment

Baghouse equipment costs consist of two components: the baghouse unit and the bags. Both costs are functions of the gross cloth area, A(gross). The

Table 5.4 Factor Method for Determining Gas/Cloth Ratio (Shaker Baghouses Only)[a]

A.

4/1 ratio		3/1 ratio		2.5/1 ratio		2/1 ratio		1.5/1 ratio	
Material	Operation*	Material	Operation*	Material	Operation*	Material	Operation*	Material	Operation*
Cardboard	1	Asbestos	1,7,8	Alumina	2,3,4,5,6	Ammonium phosphate fert.	2,3,4,5,6,7	Activated charcoal	2,4,5,6,7
Feeds	2,3,4,5,6,7	Aluminum dust	1,7,8	Carbon black	4,5,6,7	Diatomaceous earth	4,5,6,7	Carbon black	11,14
Flour	2,3,4,5,6,7	Fibrous mat'l.	1,4,7,8	Cement	3,4,5,6,7	Dry petrochem.	2,3,4,5,6,7,14	Detergents	2,4,5,6,7
Grain	2,3,4,5,6,7	Cellulose mat'l.	1,4,7,8	Coke	2,3,5,6	Dyes	2,3,4,5,6,7	Metal fumes, oxides and other solid dispersed products	10,11
Leather dust	1,7,8	Gypsum	1,3,5,6,7	Ceramic pigm.	4,5,6,7	Fly ash	10		
Tobacco	1,4,6,7	Lime (hydrated)	2,4,6,7	Clay and brick dust	2,4,6,12	Metal powders	2,3,4,5,6,7,14		
Supply air	13	Perlite	2,4,5,6	Coal	2,3,6,7,12	Plastics	2,3,4,5,6,7,14		
Wood, dust, chips	1,6,7	Rubber chem.	4,5,6,7,8	Kaolin	4,5,7	Resins	2,3,4,5,6,7,14		
		Salt	2,3,4,5,6,7	Limestone	2,3,4,5,6,7	Silicates	2,3,4,5,6,7,14		
		Sand	4,5,6,7,9,15	Rock, ore dust	2,3,4,5,6,7	Starch	6,7		
		Iron scale	1,7,8	Silica	2,3,4,5,6,7	Soaps	3,4,5,6,7		
		Soda ash	4,6,7	Sugar	3,4,5,6,7				
		Talc	3,4,5,6,7						
		Machining Operation	1,8						

*Cutting—1   Mixing—4   Conveying—7   Furnace fume—10   Intake cleaning—13
Crushing—2   Screening—5   Grinding—8   Reaction fume—11   Process—14
Pulverizing—3   Storage—6   Shakeout—9   Dumping—12   Blasting—15

Table 5.4, continued

| B. Fineness factor | | C. Dust load factor | |
Size (μm)	Factor	Loading (gr/ft^3)	Factor
>100	1.2	1–3	1.2
50–100	1.1	4–8	1.0
10–50	1.0	9–17	0.95
3–10	0.9	18–40	0.90
1–3	0.8	>40	0.85
<1	0.7		

Source: Turner, J. H., A. S. Viner, R. E. Jenkins, and W. M. Vatavuk. "Sizing and Costing of Fabric Filters, Part I: Sizing Considerations," *Journal of the Air Pollution Control Association*, June 1987, pp. 749-759. (Reprinted with permission.)

[a]This information constitutes a guide for commonly encountered situations and should not be considered a "hard-and-fast" rule. Air-to-cloth ratios are dependent on dust loading, size distribution, particle shape and "cohesiveness" of the deposited dust. These conditions must be evaluated for each application. The longer the interval between bag cleaning, the lower the air-to-cloth ratio must be. Finely divided, uniformly sized particles generally form more dense filler cakes and require lower air-to-cloth ratios than when larger particles are interspersed with the fines. Sticky, oily particles, regardless of shape or size, form dense filter cakes and require lower air-to-cloth ratios.

Table 5.5   Price Parameters for Baghouse Compartments

Baghouse Type	Correlation Range (thousand ft2)	Component	Parameter a	Parameter b
Shaker	4–16	Basic unit	3,910	7.46
(intermittent)	4–16	SS[a]	13,400	3.83
	4–16	Insulation[b]	2,080	0.50
Shaker	4–60	Basic unit	41,500	8.27
(continuous)	20–60	SS	28,500	5.45
	20–60	Insulation	0	0.371
Pulse-jet	4–16	Basic unit	10,700	6.15
(common housing)	4–16	SS	12,200	5.27
	4–16	Insulation	1,580	1.03
Pulse-jet	4–14	Basic unit	52,300	8.11
(modular)	4–16	SS	28,100	7.82
	4–16	Insulation	3,320	2.30
Reverse-air	10–80	Basic unit	32,400	7.76
	10–80	SS	15,800	6.11
	10–80	Insulation	1,250	0.886
Custom-built	100–400	Basic unit	249,000	6.11
	100–400	SS	104,000	2.56
	100–400	Insulation	66,600	0.705

Source: Turner, J. H., A. S. Viner, R. E. Jenkins, and W. M. Vatavuk. "Sizing and Costing of Fabric Filters, Part II: Costing Considerations," *Journal of the Air Pollution Control Association*, September 1987, pp. 1105-1112. These parameters, escalated to **June 1988** dollars, apply to Equation 5.3. "Basic unit" and "insulation" costs were escalated via the Chemical Engineering Plant Index ("Equipment" component); "SS" (stainless steel) costs were escalated using the Producer Price Index ("Metal Products" component).
[a]Additional cost of 304 stainless steel for metal portions of baghouse shell that contact waste gas stream.
[b]Additional cost of 3 in. of glass fiber insulation encased in a metal skin. Only that part of shell in contact with waste gas is insulated.

baghouse unit cost is, in turn, comprised of the basic baghouse cost (without bags) and the costs of "add-ons" for 304 stainless steel and fiberglass insulation. Such costs were compiled in reference 6 for six types of baghouses: intermittent shakers, continuous shakers, common-housing and modular (i.e., separate module construction) pulse-jets, reverse-air, and custom-built. All prices (basic unit or add-on) were of the form:

$$\text{Price (\$)} = a + bA(\text{gross}) \qquad (5.3)$$

where a, b = regression parameters

Values for a and b are listed in Table 5.5. Note that the prices have been escalated to **June 1988** from a third quarter 1986 base date. Also notice that the parameters apply to only certain bag area ranges, outside which they do NOT apply.

Selected bag prices (in $/ft^2$) are given in Tables 5.6 and 5.7 for pulse-jet and shaker/reverse-air filters, respectively. These prices have also been escalated to **June 1988** dollars via the "Equipment" component of the "Chemical Engineer-

Table 5.6   Selected Bag Prices for Pulse-Jet Filters

Bag Material[a]	Bag Removal Method[b]	
	TR	BBR
PE	0.47–0.64	0.35–0.40
PP	0.48–0.66	0.36–0.43
NO	1.69–2.03	1.28–1.48
HA	0.77–1.00	0.63–0.71
FG	1.17–1.40	1.03–1.34
TF	7.36–9.79	7.26–9.50

Source: Turner, J. H., A. S. Viner, R. E. Jenkins, and W. M. Vatavuk. "Sizing and Costing of Fabric Filters, Part II: Costing Considerations," *Journal of the Air Pollution Control Association*, September 1987, pp. 1105-1112. All prices are in $/ft^2, escalated to June 1988 dollars, via the Producer Price Index ("Finished Fabrics" component). The ranges in prices correspond to ranges in bag diameter.
[a]Materials: PE = 16-oz. polyester; PP = 16-oz. polypropylene; NO = 14-oz. Nomex; HA = 16-oz. homopolymeracrylic; FG = 16-oz. Fiberglas® with 10% Teflon®; TF = 22-oz. Teflon® felt. (All bags are felted except the Fiberglas®, which is woven.)
[b]TR = top bag removal (snap-in); BBR = bottom bag removal.

ing Plant" index. Table 5.7 lists reverse-air filter bag prices, both with and without sewn-in "snap rings." Similarly, shaker bag prices are shown for both "strap" and "loop" top designs.

Moreover, the pulse-jet bag prices in Table 5.6 do *not* include the prices of protective cages. Depending on the filtering area, these cages cost from $1.10 to $1.30/ft^2 of filter area (**June 1988** dollars), depending on the size and number of bags in the filter. This price range assumes mild steel cages with flanged tops and flow control venturis. Stainless steel cages would be approximately 2.5 times this amount.

Table 5.7   Selected Bag Prices for Shaker and Reverse-Air Filters

Bag Material[a]	Shaker		Reverse-Air[c]	
	Strap[b]	Loop	w/ Rings	w/o Rings
PE	0.49	0.47	0.50–0.51	0.35
PP	0.52	0.49	N.A.	N.A.
NO	1.38	1.27	1.82–1.86	1.26–1.30
HA	0.81	0.71	N.A.	N.A.
FG	N.A.	N.A.	0.82–1.07	0.57–0.75
CO	0.48	0.42	N.A.	N.A.

Source: Turner, J. H., A. S. Viner, R. E. Jenkins, and W. M. Vatavuk. "Sizing and Costing of Fabric Filters, Part II: Costing Considerations," *Journal of the Air Pollution Control Association*, September 1987, pp. 1105-1112. All prices are in $/ft^2, escalated to **June 1988** dollars, via the Producer Price Index ("Finished Fabrics" component). For the price of "Gore-Tex" bags, multiply above costs by 3 to 4.5.
[a]Materials: PE = 16-oz. polyester; PP = 16-oz. polypropylene; NO = 14-oz. Nomex; HA = 16-oz. homopolymeracrylic; FG = 16-oz. Fiberglas® with 10% Teflon®; CO = cotton. (All bags are woven.)
[b]Denotes bag top design.
[c]Prices are given for bags with and without sewn-in snap rings. Ranges shown reflect different bag diameters (i.e., 8-in. vs. 11.5-in.).

Example: Let's estimate the price of the shaker baghouse in the previous example. Assume that the waste gas flowrate is 21,500 acfm at 100°F with a 2% moisture content.

Solution:

Cloth area: In the previous example, we calculated a gas/cloth ratio of 2.4. This yields a net (filtering) cloth area of 21,500 acfm/2.4 = 8,960 ft². The "f" factor in Table 5.2 corresponding to this net area is 1.50, so that:

$$A(gross) = 8,960 \times 1.50 = 13,400 \ ft^2$$

Baghouse shell: Because the moisture content and temperature are both moderate, no insulation will be needed. Also, as the waste gas is noncorrosive, carbon steel construction will suffice. Therefore, from Table 5.5, we select the following parameters: a = 41,500; b = 8.27. Thus:

$$Shell \ price = 41,500 + 8.27(13,400 \ ft^2) = \$152,300$$

Bags: Because the gas temperature is less than 180°F, we can use cotton bags. The price for these (Table 5.7) is $0.48/ft², assuming "strap top" design. Hence:

$$Bag \ cost = \$0.48/ft^2 \times 13,400 \ ft^2 = \$6,400$$

Finally, the sum of the baghouse shell and bag prices is:

$$\$152,300 + \$6,400 = \$159,000 \ (rounded)$$

Total Capital Investment (TCI)

As the Chapter 2 costing procedure shows, the TCI is the product of the *direct-indirect installation factor* and the *purchased equipment cost* (PE). As Table 2.2 shows, the average composite installation factor for fabric filters is 2.17. The PE, in turn, is the sum of the costs of the baghouse and bags, auxiliary equipment, sales taxes, freight, and instrumentation. Auxiliary equipment usually includes a fan and motor, screw conveyor, and stack. It may also include a cyclone for lightening the inlet dust loading. For auxiliary equipment costs, see Chapter 4.

Annual Costs

These consist of *direct* and *indirect* annual costs and *recovery credits*. The *direct* costs include operating and supervisory labor, maintenance labor and materials, electricity, compressed air, dust disposal, and replacement bags. The unit costs of these items will vary from source to source. However, we can present general consumption figures:[7]

- operating labor: 2 to 4 hours/shift
- supervisory labor: 15% of operating labor
- maintenance labor: 1 to 2 hours/shift
- maintenance materials: 100% of maintenance labor
- electricity:
  - system fan: 5- to 10-in. pressure drop
  - reverse-air fan: 6- to 7-in. pressure drop
  (Calculate power consumption for either fan via Equation 2.2.)
  - shaker mechanism: $6.05 \times 10^{-6}A$ (gross) (kW)
- compressed air (pulse-jet filters): 2 scf/1000 acfm
- dust disposal:
  - landfill fee: $10 to $40/ton (nonhazardous dust)
    $250/ton (RCRA-hazardous dust)
  - hauling cost: $0.05 to $0.20/yd^3-mi (see "Wet Scrubbers," below)
- bag replacement: As discussed in Chapter 2, treat the bags as an investment amortized over their useful life (1 to 5 years; 2 years, typical).

Thus, bag replacement is calculated as:

$$C_{br} = (P_b + P_{bl})CRF_b \qquad (5.4)$$

where $\quad P_b =$ price of full set of bags, including taxes and freight ($)

$\qquad P_{bl} =$ labor cost to replace bags (10 to 20 and 5 to 10 man-min/bag for reverse air/shakers and pulse-jets, respectively)

$\qquad CRF_b =$ capital recovery factor for useful life of bags and annual interest rate (see Eq. 2.4)

As shown below, this bag replacement cost must be deducted from the TCI before calculating the annual capital recovery cost.

The *indirect annual costs* consist of overhead, capital recovery, property taxes, insurance, and administrative charges. As Chapter 2 shows, the last three items total 4% of the TCI. The *overhead* is figured as 60% of the sum of operating, supervisory, and maintenance labor and maintenance materials, as Chapter 2 suggests. The capital recovery cost (CRC) is calculated as follows:

$$CRC = (TCI-P_b-P_{bl})CRF_s \qquad (5.5)$$

where $\quad CRF_s =$ capital recovery factor for the baghouse (excluding bags) and rest of the control system.

As Chapter 2 shows, the useful life of a baghouse system ranges from 5 to 40 years, with 20 years being typical.

Lastly, *recovery credits* would be considered if the recovered dust could be reused in the process or sold. To calculate this credit, simply multiply the amount of dust reused/sold by its assigned value. This credit is then subtracted from the sum of the direct and indirect annual costs to yield the net *total annual cost (TAC)*.

ELECTROSTATIC PRECIPITATORS

Description

Electrostatic precipitators (ESPs) are particle control devices that employ electrical forces to remove particles from a moving waste gas stream onto collecting plates. This removal is effected by charging the particles by passing them through gaseous ions that have, in turn, been charged by high-voltage electrodes located between the plates. After a time, the plates are knocked ("rapped") and the accumulated particles fall into collecting hoppers. In certain ESP designs, the plates are cleaned with wash water.

The five most commonly used ESP designs are: *plate-wire, flat plate, tubular, wet,* and *two-stage*.[8] In *plate-wire* ESPs, the most prevalent design, the waste gas flows between parallel sheet metal plates and high-voltage electrodes — long, weighted wires hanging between the plates or supported by rigid frames. The high voltage they generate (20,000 to 100,000 v.) is provided by a step-up transformer, high-voltage rectifiers, and, occasionally, filter capacitors. Because these ESPs permit parallel gas flow through several sections of plates, they are suited for large gas flowrates.

Plate-wire (and other) ESPs have two weaknesses: first, during cleaning ("rapping"), some of the collected particles become reentrained in the waste gas and pass out of the unit. For coal fly ash, this loss can exceed 10%. Also, if the resistivity of the collected particle layer is too high (above 2×10^{11} ohm-cm) the charging field breaks down and sparking ("back corona") may occur. This also impairs the collection efficiency. Figure 5.2 depicts the components of a typical ESP.

Flat-plate ESPs are used because they provide both greater collecting plate surface and a higher charging field. The charging is accomplished by electrodes (needles or wires) installed ahead of or behind the collecting plate zones. This configuration minimizes back corona problems, but may increase reentrainment losses. Flat-plate ESPs are used to collect small (1- to 2-μm) high-resistivity particles from moderately large (100,000- to 200,000-acfm) gas streams.

In *tubular* precipitators, the high-voltage electrode is placed inside a tube (circular, square, or hexagonal) that serves as the collecting surface. Generally used on heavy industry sources (e.g., iron and steel plants), tubular ESPs are one-stage units in which all the gas passes through the electrode region.

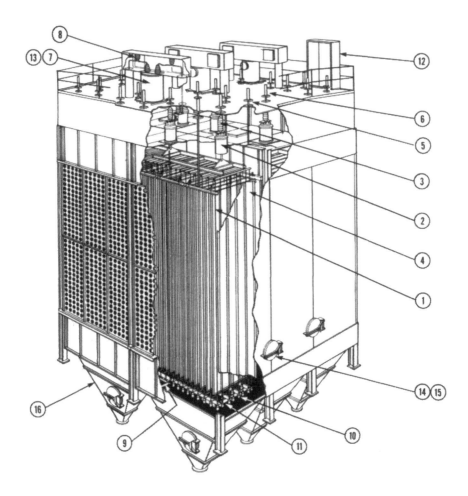

(1) High-voltage discharge electrodes (weighted wire or rigid electrode)
(2) HV support insulators (porcelain or alumina)
(3) HV rapper or vibrator insulators
(4) Grounded collecting electrodes
(5) Grounded electrode rappers or vibrators
(6) High-voltage rappers or vibrators
(7) Transformer-rectifier set components
(8) Bus duct insulators
(9) Lower HV anti-sway insulators
(10) HV discharge electrode weights
(11) Lower HV steadying frames
(12) Rapper and vibrator control components
(13) Transformer-rectifier control components
(14) Access doors
(15) Key interlock systems
(16) Collecting hoppers

Figure 5.2 Components of an electrostatic precipitator (courtesy Field Service Associates, Inc.).

Because the particles they collect are typically wet or sticky, these units are cleaned with water. Thus, they are *wet ESPs*.

In fact, *wet precipitators* may be of any of the above designs. Their distinguishing feature is that the collecting plates are cleaned with water, instead of being rapped. Neither dust reentrainment nor back corona occur in wet ESPs. However, the wash water composition must be carefully controlled, and the wastewater has to be treated before ultimate disposal.

In another type of wet ESP, the *two-stage*, the discharge electrode (ionizer) and collecting surface are in separate compartments, the former preceding the latter. This design affords several advantages—more particle charging time, fewer back corona problems, and economical construction for small sizes. Generally used for smaller (50,000 acfm or less) gas flows, two-stage ESPs are often used to collect sticky sub-micrometer particles (oil mists, fumes, etc.). They are cleaned by a detergent water wash followed by air-blow drying. Because they are smaller than the other types, two-stage ESPs are usually sold as packaged systems.

To facilitate their operation, ESPs require certain auxiliary equipment. This includes ductwork, preconditioners (e.g., cyclone), fan-and-motor, and stack. In addition, wet ESPs require equipment (e.g., pumps) to supply, store, inject, and recirculate the wash water. Finally, to lower dust resistivity to improve dry ESP performance, the waste gas is sometimes conditioned by injecting sulfur trioxide (10 to 30 ppm, typically). Conditioning equipment usually includes a molten sulfur storage vessel, oxidizing burner, catalyst, and dust injection probes. Other chemicals, such as ammonia, may also be added to the waste gas to improve collection.

Along with the aforementioned dust resistivity, several other variables affect ESP performance.[9] These are:

- Particle inlet loading: Because the ESP is a constant-efficiency device, the inlet/outlet loading ratio remains constant over a wide range. At excessively high loadings, however, the ESP operation will be adversely affected.

- Particle size distribution: Usually, an ESP will collect all particles > 10 μm in diameter, so that the distribution of interest is ≤ 10 μm. To describe a particle size distribution, the two parameters needed are the *mass median diameter (MMD)* and the *geometric standard deviation*. (*Note:* These parameters are described in the "Wet Scrubbers" section below.)

- Gas characteristics: These include the gas *volumetric flow rate*, *temperature*, and *composition* (especially moisture). The last two parameters can strongly affect the resistivity of the captured dust. Gas characteristics can also influence the size of the ESP power supplies.

Lastly, a few words about construction.[10] Most ESP shells are constructed of 3/16-in. or 1/4-in. carbon steel plate, while the collecting electrodes (plates) are commonly 18-gauge. Discharge wires, also carbon steel, range in diameter from 0.1 to 0.25 in. Except in wet ESPs, stainless steel is rarely used, because it is often prone to fatigue failure in dry ESPs. However, in special applications, different materials of construction (such as lead-lined steel) may be used.

Sizing Procedure

The most commonly used expression for sizing electrostatic precipitators is the well-known Deutsch-Anderson equation:

$$E = 1 - \exp(-v_m A/Q) \qquad (5.6)$$

where E = overall particle collection efficiency (fractional)
v_m = effective migration velocity (ft/min)
A = precipitator collecting area (ft^2)
Q = waste gas flow rate (acfm)

Moreover, by definition:

$$A/Q = \text{specific collecting area ("SCA," min/ft)}$$

We must emphasize that the Deutsch-Anderson equation provides a mean collection efficiency, weighted according to the size distribution of the inlet particles. Clearly, smaller particles would be captured less efficiently than larger. This efficiency-size relationship is demonstrated mathematically in references 11 and 12. The migration velocity serves as the weighting parameter in Equation 5.6.

Values of the migration velocity (v_m) are listed in Table 5.8 for three ESP types, selected particle sources, and a range of overall collection efficiencies. Notice that v_m varies with the efficiency in some cases, but is constant in others. These migration velocities were calculated via the complex (and relatively new) Lawless-Sparks procedure, as described in reference 11. In this procedure, the migration velocity varies not only with particle size, but also with the fractions of reentrainment and "sneakage" (i.e., particles that slip through the ESP without being collected), and the electric field strength. Because the last three parameters vary with the ESP type, we see the velocity differences that appear in Table 5.8.

Example: If the above cement plant ball mill emissions were controlled by a 99.5% efficient plate-wire ESP instead of a baghouse, how much collecting area would it require?

Solution: From Table 5.8, we find a value for a cement kiln dust, but nothing for a ball mill. Further, this value (0.059 ft/sec. = 3.5 ft/min) was calculated for a 600°F stream temperature, versus 100°F for the ball mill offgas. Can we use it? Yes—as long as we realize that the result will

Table 5.8 Selected Migration Velocities

Particle Source[a]	ESP Type[b]	Design Efficiency (%)			
		95	99	99.5	99.9
Bituminous coal	P-W (dry)	0.41	0.33	0.31	0.27
fly ash (300)[c]	P-W (wet)	1.03	1.08	1.11	0.82
	F-P	0.43	0.50	0.61	0.53
Sub-bituminous	P-W (dry)	0.56	0.39	0.34	0.29
coal fly ash	P-W (wet)	1.31	1.40	1.45	1.03
(300)	F-P	0.94	0.60	0.70	0.58
Other coal	P-W (dry)	0.32	0.26	0.26	0.24
(300)	P-W (wet)	0.69	0.70	0.71	0.56
	F-P	0.51	0.37	0.50	0.44
Cement kiln	P-W (dry)	0.049	0.049	0.059	0.059
(600)	P-W (wet)	0.21	0.18	0.16	0.19
	F-P	0.079	0.075	0.11	0.10
Glass plant	P-W (dry)	0.052	0.052	0.049	0.049
(500)	P-W (wet)	0.15	0.15	0.14	0.13
	F-P	0.059	0.062	0.085	0.085
Iron/steel sinter	P-W (dry)	0.20	0.20	0.22	0.21
plant (w/mechanical	P-W (wet)	0.46	0.45	0.44	0.38
collector)	F-P	0.44	0.40	0.43	0.41

Source: Turner, J. H., P. A. Lawless, T. Yamamoto, D. W. Coy, G. P. Greiner, J. D. McKenna, and W. M. Vatavuk, "Sizing and Costing of Electrostatic Precipitators, Part I: Sizing Considerations," *Journal of the Air Pollution Control Association*, April 1988, pp. 458-471. These migration velocities (in ft/sec) have been calculated via the Lawless-Sparks procedure described in this source.
[a]In all cases, *no* back corona has been included.
[b]P-W (dry), P-W (wet) = plate-wire (dry) and plate-wire (wetted wall); F-P = flat-plate designs.
[c]Denotes precipitation temperature (°F).

be approximate and that in particular it will probably undersize the ESP. (ESPs are typically less efficient at lower operating temperatures.) That said, let's compute the collecting area. First, rearrange Equation 5.6, to solve for A:

$$A = (-Q/v_m)\ln(1 - E) \qquad (5.7)$$

Substituting, we obtain:

$$A(ft^2) = (-21,500/3.5)\ln(1 - 0.995) = 33,000 \text{ ft}^2$$

(Notice that we've rounded the collecting area to two figures, to be consistent with the two-place precision of the migration velocity.)

Costing Procedure

Equipment Costs

These costs are almost always correlated with the collecting area. As the preceding section showed, the area encapsulates the many variables that affect the performance of an ESP. With this in mind, the authors of reference 13 obtained cost quotes from ESP vendors and regressed them against their collecting areas. These quotes fell into two area ranges: 10,000 to 50,000 ft² and 50,000 to 1,000,000 ft². Except for those units with plate areas under 15,000 ft², all ESPs were field-erected. Escalated to **June 1988** dollars (from a second-quarter 1987 base), via the *Chemical Engineering* Plant Index ("Equipment" component), these costs are:

$$\text{Price (\$)} = aA^b \tag{5.8}$$

where A = collecting area (ft²)
          a, b = regression parameters

The values for a and b are:

Plate Area (1000 ft²)	a	b
10 to 50	962	0.628
50 to 1000	90.6	0.843

These costs ($\pm 25\%$ accurate) apply to all ESP designs, except the two-stage. The prices include the following:

- ESP casing
- pyramidal hoppers
- rigid electrodes and internal collecting plates
- transformer-rectifier sets and microprocessor controls
- rappers and stub-supports (legs)

Also included in the prices are such "standard options" as inlet and outlet nozzles and diffuser plates; hopper auxiliaries/heaters and level detectors; weather enclosure and stair access; structural supports; and insulation (3 in. of fiberglass encased in a metal skin). These options comprise about 1/3 of the above costs.

Although Equation 5.8 provides costs for wet ESPs in typical applications, prices for these units placed in severe service would be considerably higher. For example, *installed* costs for wet ESPs controlling acid mist range from $70 to $130/ft² of collecting area. To control coke oven offgas, the cost range would be comparable: $100 to $130/ft².

Because they are packaged units, two-stage precipitators are costed differently from their larger counterparts. Estimate their cost (also in **June 1988** dollars) from the following equation:[14]

$$\text{Price (\$)} = -276 + 39.0\ln Q \tag{5.9}$$

where Q = gas flow rate (acfm)

These costs include a two-cell ESP unit, inlet and outlet plenums, prefilter, cooling coils with coating, coil plenums with access, water flow controls, triple pass configuration, system exhaust fan with accessories, and in-place foam cleaning system with semiautomatic controls and programmable controller—all mounted on a steel structural skid.

Example: Estimate the equipment cost for the plate-wire ESP sized in the previous example.

Solution: Recall that we calculated a plate area of 33,000 ft^2. This falls in the lower area range of Equation 5.8. The price is, therefore:

$$P(\$) = 962(33,000)^{0.628} = \$660,000$$

Total Capital Investment

As with fabric filters, the TCI is the product of the composite installation factor and the purchased equipment cost. From Table 2.2, we obtain an average composite installation factor of 2.24.

This factor, however, applies to all but the two-stage ESPs. Because the two-stage precipitators are packaged units, a lower installation factor should be used. A factor of 1.2 to 1.3 times the ESP purchase price (including freight and sales taxes) should provide an adequate estimate of the TCI.

Annual Costs

Direct annual costs include those for operating and supervisory labor, maintenance labor and materials, electricity, wash water, and dust disposal/wastewater treatment. Unit prices will vary, but consumption figures can be given (courtesy of reference 15):

- Operating labor: 0.5 to 2 hours/shift
- Supervisory labor: 15% of operating labor
- Maintenance labor
  - for A > 50,000 ft^2: 0.15 hr/wk-1000 ft^2
  - for A ≤ 50,000 ft^2: 8 hr/wk
- Maintenance materials: 1% of ESP *equipment* cost
- Electricity
  - system fan: 0.1 to 0.4-in. pressure drop (calculate power consumption via Equation 2.2)
  - transformer-rectifier set and rapper system: 0.00194A (kW)
  - two-stage ESPs: 0.025 to 0.100 kW/1000 acfm (0.040, typical)
- Wash water
  - large wet ESPs: 5 gpm/1000 acfm
  - two-stage ESPs: 16 gpm/1000 acfm
- Dust disposal: see "Fabric Filters" section
- Wastewater treatment: see "Wet Scrubbers" section

Procedures for calculating the *indirect annual costs* and *recovery credits* are the same as those presented in the "Fabric Filters" section. The only difference is, the ESP capital recovery cost (CRC) is calculated using the *entire* TCI, as no replacement parts costs apply here.

WET SCRUBBERS

As we've seen in the previous two sections, several different electrostatic precipitator and fabric filter designs are used to control air emissions. With wet scrubbers, however, "several" doesn't begin to describe the many choices available to the user. There are literally dozens of designs and design variations to choose from, each offering "unique" features and promising "minimal" operating and maintenance costs. Few authors have succeeded in distilling the essence of this complex subject from the proliferation of literature available these days. Schifftner and Hesketh are notable exceptions. In *Wet Scrubbers: A Practical Handbook* (Chelsea, MI: Lewis Publishers, Inc., 1986), they give us a concise but comprehensive treatment on the applications, design, and operation/maintenance of wet scrubbers. We're privileged to use it as the basis of the "Description" and "Sizing Procedure" portions of this section.

Description

Unlike most other control devices, wet scrubbers can (and usually do) collect both particulate and gaseous emissions. Some scrubber types are better suited to collect the one better than the other, however. Thus, we conveniently categorize scrubbers as being either particulate or gaseous collectors. *Particulate* collectors employ inertial or (in rare cases) electrostatic forces to capture emissions. They consist of particle scrubbers and mist eliminators. *Gaseous* control devices collect gases by absorbing them with liquids in units having large liquid-gas contact areas. Further, the pressure drop and other physical forces are minimized in gaseous scrubbers, but maximized in particulate collectors.[16]

Particle Scrubbers

Most particle scrubbers rely on inertial (Newtonian) forces to capture emissions. These forces are (predominantly): (1) impaction, (2) interception, and (3) diffusion. In most particle scrubbers, all three phenomona occur to a greater or lesser extent. With *impaction*, the most prevalent mechanism, contaminants are ". . .accelerated and impacted onto a surface or into a liquid droplet."[17] The particle's kinetic energy is used to penetrate the surface tension of a scrubbing liquid droplet, allowing it to surround the particle, thereby increasing the droplet's density so that it can be more easily separated inertially. Impaction devices include *venturis* and *spray chambers*.

While in impaction the particles and droplets collide, in *interception* they merge like traffic at a freeway interchange. Here, the particles flow nearly parallel to (instead of perpendicular to) the droplets. Interception works best on sub-micrometer particles which, given their low mass, tend to follow the gas flow lines. *Spray-augmented scrubbers* and *high-energy venturis* employ this mechanism.

Diffusion is operative with particles smaller than 0.5 μm and where there is a high temperature difference between the gas and scrubbing liquid. The particles migrate through the spray along lines of irregular gas density and turbulence, contacting droplets of approximately equal energy. With very small (<0.05 μm) particles, intermolecular interaction occurs (*diffusiophoresis* and/or *thermophoresis*). However, neither phenomenon is critical in scrubber design.

Six particle scrubber designs are used in control applications:[18] (1) spray, (2) wet dynamic, (3) cyclonic spray, (4) impactor, (5) venturi, and (6) augmented. In all these scrubbers, impaction is the main collection mechanism for particles larger than 3 μm. Because smaller particles respond to noninertial capture, designers try to create a high density of small liquid droplets in the scrubber contact zone to trap the particles. This is done at the price of a high energy consumption due to hydraulic or velocity pressure losses.

Spray scrubbers (also known as "spray chambers") use atomized streams of scrubbing liquid to remove particulates via direct impaction or interception on liquid droplets or diffusion of the particles to these droplets. The spray nozzles are installed on headers or mounted on the walls of horizontal or vertical vessels. Because of the large contact surface area the sprays create, these scrubbers also can be used as gas absorbers. Typical capture efficiencies for various particle sizes are: >5 μm: 90%; 3 to 5 μm: 60 to 80%; sub-micrometer: 40 to 50%.

Wet dynamic scrubbers (also known as "wetted fans") employ sprays to introduce liquid to the inlet of a paddle wheel fan, in which the particles are removed by impaction on the droplets. The dirty droplets are caught in a droplet eliminator, while the cleaned gas passes through. Though cheap to purchase and operate, dynamic scrubbers are not very efficient — i.e., only 60 to 75% on sub-micrometer particles — and are quite sensitive to abrasive dusts.

Cyclonic spray scrubbers are similar to dynamic scrubbers, employing the spray nozzles at the inlet of a cyclonic separator instead of a fan wheel. The gas enters tangentially a cylindrical vessel that is tapered to increase the gas velocity and droplet density to effect particle capture by impaction and interception. Because spray nozzles are used, the recirculating scrubber liquor must be cleaned via strainers and/or decanters that normally come with the unit. Finally, without the spray nozzles, a cyclonic scrubber becomes a *separator*, which can be used after any scrubber (e.g., a venturi) to remove entrained liquid droplets.

Both cyclonic and dynamic scrubbers have similar removal efficiencies: >5-μm particles: 95%; sub-micrometer: 60 to 75%.

Impactor ("impingement") scrubbers come in many forms. In each, the scrubbing liquid flows over flat trays, while the particle-laden gas flows up through them. The trays may be perforated with overlaid "baffle strips," moveable valve

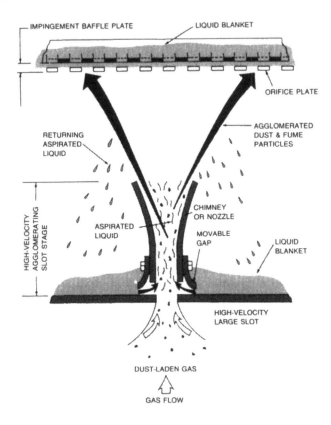

Figure 5.3 Adjustable slot plate impingement scrubber (courtesy Flakt/Peabody Division).

discs, bubble caps, or adjustable impingement trays. The particle capture occurs when the waste gas, accelerated to a very high velocity, is directed against a "target plate" cleaned by flowing scrubbing liquid. (See Figure 5.3.) However, these scrubbers tend to plug at high particle loadings (around 10 grains/scf) and require dilute scrubbing liquids (<2% solids by weight).

Venturi scrubbers are best suited for removing sub-micrometer particles. Venturis employ shearing and impaction forces to atomize water, producing a high-density distribution of fine droplets. These droplets collide with the particles immediately upstream of the venturi "throat." This throat is a converging section through which the gas is conveyed downward to accelerate it to high velocities (450 ft/sec or more). Throats can be "fixed" (round or rectangular) or "adjustable." The latter can be constricted/dilated to increase/decrease the gas ("throat") velocity. All of this comes at a steep price, for gas pressure losses in venturis can range up to over 100 in. w.c., depending on the particle size distribution and desired removal efficiency. (See "Sizing Procedure" for more on this.)

Downstream of the throat, the gas velocity decreases as it passes through the "diverging" section of the venturi. The gas stream next makes a 90-degree turn

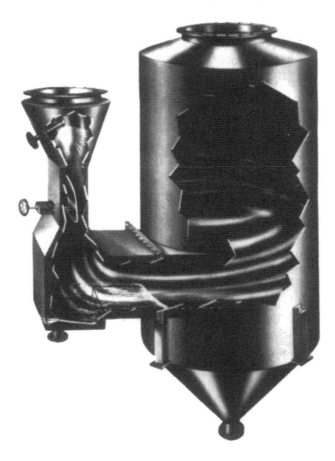

Figure 5.4   Peabody venturi scrubber with cyclonic demister (courtesy Flakt/Peabody Division).

through a "flooded elbow" containing scrubbing liquid that traps larger droplets. From here, the gas flows to a cyclonic separator for removal of mist and other droplets not caught by the elbow, and thence to the stack. (See Figure 5.4.) Meanwhile, the spent scrubber liquor usually flows to a recirculation tank, whence a fraction of the liquid is "bled" from the system to keep the solids concentration limited to 20 to 30% by weight. The rest of the liquid is recycled to the scrubber. Finally, makeup water is added to the recirculation tank to compensate for the bleed and evaporative losses.

Variations in the basic venturi design include the *annular* venturi (an adjustable throat type), various *eductor* designs, and electrostatic charge–enhanced units. A venturi provides the ultimate design for removing sub-micrometer particles via purely mechanical means. To go beyond it, other forces have to be employed, most prominently electrostatic attraction. In *electrostatically enhanced augmented* scrubbers, a positive charge is applied to the incoming stream, causing the particles to be attracted to the electronegative water droplets. But due to the polarity of the water molecule, there have been developmental difficulties with augmented scrubbers. Hence, their applications have been limited.

Mist Eliminators

Also called *entrainment separators* and *demisters*, mist eliminators serve to remove 99 to 99.9% of liquid droplets from the scrubber exit stream. They employ inertial impaction, centrifugal force, and even diffusion and interception capture mechanisms. Most mist eliminators are of either *baffle* or *mesh* design and employ either horizontal or vertical gas flow.[19]

Horizontal units permit higher velocities and liquid loadings, because the separated liquid droplets do not fall back into the incoming gas, as they do in vertical units. They often include liquid phase separation chambers to reduce liquid carry-through. Vertical-flow eliminators typically use the *Chevron*-type multipass (2-6) baffles. The baffles may be either "zigzag" or "slanted" in design.

Mesh-type eliminators are used for vertical gas flow when no sticky particles are present that might plug the mesh. (See Figure 5.5.) The mesh pads are from 4 to 12 in. thick (6 in. is typical) and may be installed 0 to 45 degrees from the horizontal. Cylindrical mesh fiber packs and packed fiber bed eliminators are also used, but rarely so in industrial applications.

Finally, all impaction-type mist eliminators require washing to keep them clean and remove buildup. A wash rate of 3 gpm/ft^2 of surface area is recommended.

Gaseous Collectors

All of the above-described particulate scrubbers also remove gases to some extent, especially those with higher liquid/gas ratios. However, the high gas velocities that afford highly efficient particulate capture are not conducive to gas absorption, which requires longer contact times. Gas absorption is, of course, a subject unto itself — one that many textbooks have treated exhaustively. Briefly, absorption is a mass transfer process in which the gaseous pollutants are transferred to the absorbent liquid. The rate of this mass transfer depends on the *liquid-gas interface area*, the *driving force* (differences in pollutant concentration between the gas and liquid phases), and the *mass transfer coefficient*, which depends on the characteristics of the absorbent, absorbate, and the liquid/gas contacting medium. The medium — trays, packing material, grids, etc. — affords intimate gas-liquid contact by maximizing the available surface area. But to maximize surface area, one must minimize the size of the packing, tray openings, etc., which causes the pressure drop to increase. Thus, there is a cost-efficiency tradeoff.

Gas absorption typically occurs in vertical towers, wherein the gas and

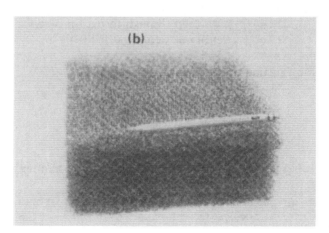

Figure 5.5 (a) Typical mesh pad mist eliminator with supports. (b) Stainless steel wire mesh section (6-in. thickness). (Reprinted from *Wet Scrubbers: A Practical Handbook* by K. C. Schifftner and H. E. Hesketh, Lewis Publishers, Inc., 1986.)

liquid flow countercurrently. Horizontal ("crossflow") scrubbers are also used. Needless to say, both packed and tray-type absorbers are subject to plugging, so that the inlet particulate loading must be kept low. For this reason, absorbers are *not* recommended for use as particulate scrubbers.

Sizing Procedure

The primary sizing parameter for *all* scrubbers—particulate, gaseous, and mist eliminators—is *gas velocity*. Notice that a "the" does not appear before "gas velocity." That is because for many scrubber designs, the velocity varies from the inlet to the outlet, so that no single value predominates. In general, gas velocities may be divided into two categories: *equipment* (e.g., entry and exit) and *removal* (e.g., impaction, centrifugal, and venturi throat velocities). For a given gas volumetric flowrate, these velocities determine the required cross-sectional area(s) of the scrubber. And the area, along with the height and wall thickness of the scrubber and the materials of construction, are the major factors in determining its price. Although guidelines for equipment velocities may be given, *removal* velocities are more difficult to determine—especially for a venturi. The primary parameter for a venturi is the *throat velocity* (ft/sec), which depends upon the pressure drop, volumetric flowrate, gas density, liquid rate, and a constant, C, as follows:[20]

$$v_t = Q/A = C(\Delta P/r_g)^{0.5} \qquad (5.10)$$

where Q = maximum volumetric flowrate (acfm)
   A = throat area (ft^2)
   ΔP = pressure drop through throat (in. w.c.)
   r_g = gas density at saturation (lb/ft^3)

In turn, C is a function of the liquid/gas ratio (L/G):

$$C = 1060\exp(-0.0279L/G) \qquad (5.11)$$

where $0 \leq L/G$, gal/1000 ft^3 gas ≤ 40

This relationship applies to rectangular throat venturis treating gas with a density of 0.06 lb./ft^3, and is accurate to within ±4%.

Table 5.9 lists equipment and removal velocities for other types of scrubbers. Where velocity ranges are given, the values vary as functions of the L/G ratio, the pressure drop, or other parameters. But how do we determine the *optimum* removal velocity? There is no cut-and-dried answer. If we increase the velocity, we decrease the scrubber cross-sectional area for a given volumetric flowrate. This allows us to build a smaller—and less expensive—scrubber. But a higher velocity produces a much higher pressure drop. As Equation 5.10 shows, the pressure drop increases with the square of the velocity. This drives up the elec-

Table 5.9   Selected Scrubber Equipment and Removal Efficiencies

Velocity Category	Scrubber Type	Velocity (ft/sec)
Equipment		
Inlet	All	45–80
	Cyclonic	90–120
Outlet	All	30–40 (w/ stack)
	All	50–60 (w/o stack)
Headers[a]	All	6–10
Drains[a]	All	1–3
Particulate Removal	Mesh eliminator	14
	Chevron eliminator	9–14 (vertical)
		15–20 (horizontal)
	Packed tower	2–6 (vertical)
		4–8 (horizontal)
	Cyclonic	105–140[b]
	Venturi	90–400[c]
	Impingement	<14
	Spray tower	10

Source: Schifftner, K. C., and H. E. Hesketh. *Wet Scrubbers: A Practical Handbook*. Chelsea, MI: Lewis Publishers, Inc., 1986, pp. 10, 24–29.
[a]*Liquid* velocities. All others gaseous velocities.
[b]Varies with pressure drop (4 to 6 in.) and gas density (0.05 to 0.068 lb/ft^3).
[c]Refer to Equations 5.10 and 5.11.

tricity consumption. So the selection of the optimum velocity embodies a trade-off between capital and operating costs. This trade-off exists for both particulate and gaseous scrubber categories. However, because the principles for sizing each category are different they will be presented separately.

Particulate Scrubbers and Mist Eliminators

One of the most succinct and comprehensive particulate scrubber selection and sizing procedures developed is the "Calvert Cut/Power Method."[21] This method consists of these steps:

1. *Determine the mass median particle diameter (d$_{50}$)*. This is done by plotting the cumulative size distribution on log-probability paper. The size distribution is measured with a particle sizing instrument (e.g., a cascade impactor). These sizes must be the *aerodynamic* ("inertial") diameters, because as we've seen, scrubbers collect particles primarily by inertial mechanisms. Depending on the particle density, the aerodynamic diameter is from 1 to 4 times the *physical* particle diameter, this ratio increasing with the density. Also, log-probability paper is used because most particle emission streams are log-normally distributed. The cumulative distribution will be a straight line on log-probability paper.

2. *Calculate the overall mass penetration (p$_t$)*. The penetration is the inverse of the collection efficiency, or:

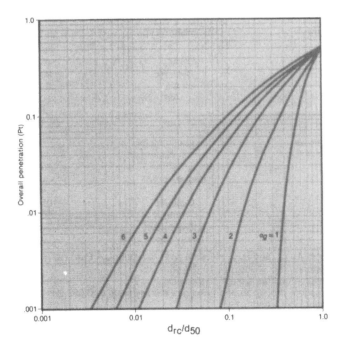

Figure 5.6 Overall penetration plot (reprinted with permission of Calvert Environmental Equipment, Inc.).

$$p_t = 1 - \text{Efficiency}/100 \qquad (5.12)$$

3. *Determine the particle cut diameter (d_{rc}).* As Figure 5.6 shows, the ratio of the cut diameter to the mass median diameter is a function of the overall penetration and the *standard deviation* (s.d.) of the particle size distribution. By definition, for log-normal distributions:

$$\text{s.d.} = d_{84}/d_{50} \qquad (5.13)$$

4. *Select the scrubber and determine its pressure drop and horsepower.* This is really three steps in one. First, as Figure 5.7 shows, more than one scrubber may be selected for a given aerodynamic cut diameter. This figure lists curves for six: packed column, sieve-plate column, fibrous packed bed, Calvert Double Scrubber™, venturi scrubber, and Collision Scrubber™. Except for the Double Scrubber™, both pressure drops and power requirements are given. The power requirement is based on an overall fan efficiency of 50%. (To obtain power requirements at different efficiencies, simply multiply the Figure 5.7 data by $50/n'$, where "n'" is the fan efficiency desired.)

If several scrubbers can do the job, we'd usually select the one with the lowest pressure drop. However, there may be cases where the capital costs of

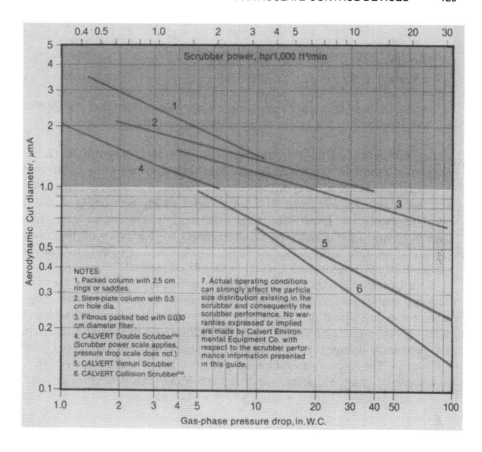

Figure 5.7 Cut/power relationship (reprinted with permission of Calvert Environmental Equipment, Inc.).

the scrubbers would also affect the selection, so that pressure drop may not be the sole consideration.

Example: The mass median (d_{50}) and 84th percentile (d_{84}) aerodynamic diameters of a particle size distribution are 1.7 and 3.4 μm, respectively. The inlet and desired outlet dust loadings are 3.0 and 0.15 grains/dscf, in turn. Select the "right" scrubber for this application.

Solution: From Equation 5.13, s.d. = 3.4/1.7 = 2.0. And, p_t = Outlet loading/Inlet loading = 0.050. Reading Figure 5.6, we obtain a d_{rc}/d_{50} ratio of 0.26.

Thus: d_{rc} = (1.7)(0.26) = 0.44 μm. Finally, as Figure 5.7 shows, two scrubbers can accommodate this cut size: a Collision Scrubber™ and a venturi. Their pressure drops and power requirements are:

Scrubber	Pressure Drop (in. w.c)	Power Requirement (hp/1000 acfm)
Collision™	17	5.4
Venturi	24	7.6

On the basis of the power requirement alone, we would select the Collision Scrubber™. But again, other costs may influence this decision.

One parameter we have yet to discuss is the liquid/gas ratio, L/G. It is especially important in venturi design, for L/Gs here can be quite high. As Equation 5.5 indicates, L/G influences the constant, C, which, in turn, affects the throat velocity and area. This ratio varies widely between (and within) scrubber categories. Experience usually dictates the selection of L/G. Table 5.10 lists typical L/Gs and pressure drop ranges for selected scrubber categories. The pressure drop values in Table 5.10 are meant to supplement, not supplant, those given by the Calvert "Cut Power Method."

Note that the L/G ratio determines the *recirculation* rate, not the water makeup requirement. As stated above, the makeup rate is the sum of the evaporative losses and the "bleed" rate. The evaporation rate depends on the waste gas inlet moisture content and temperature and the scrubber operating temperature—usually (but not always) the temperature of the gas at 100% saturation. The first two variables are given, while the last is obtained from a psychometric chart. (Schifftner and Hesketh discuss this procedure in detail.)

The bleed rate (g_b, gpm) is straightforward to calculate:

$$g_b = 0.120 M_{in} E/s \qquad (5.14)$$

where M_{in} = inlet particulate loading (lb/min)
E = overall control efficiency (fractional)
s = allowable solids content in recirculating scrubber liquid (lb/lb water)

For venturis, $s = 0.20$ to 0.30, while for spray scrubbers it is much lower—0.005 to 0.01.

Table 5.10 Liquid/Gas (L/G) Ratios and Pressure Drops for Selected Scrubbers

Scrubber	Liquid/Gas Ratio (gal/1,000 acfm)	Pressure Drop (in. w.c.)
Spray	—	2–4
Impactor (impingement)	2–10	2–3[a]
Venturi	4–100	< 100
Cyclonic	7	4–6
Mist eliminators	3[b]	0.2–1.5

Source: Schifftner, K. C., and H. E. Hesketh. *Wet Scrubbers: A Practical Handbook*. Chelsea, MI: Lewis Publishers, Inc., 1986, pp. 10, 24–29.
[a]Pressure drop *per scrubber tray*.
[b]Liquid rate measured in gpm/ft² of impaction-type eliminator surface area.

Example: The waste gas in the previous example has an inlet volume flowrate of 10,000 acfm, a temperature of 300°F, and a 20% moisture content. Recall that its dust loading and control efficiency were 3.0 grains/dscf and 95%, respectively. If the gas reaches saturation in the scrubber, what will be its final volume flowrate and temperature and the water make-up rate?

Solution: At this moisture content, the water vapor and dry air flowrates are 2000 and 8000 acfm, respectively. The corresponding flowrates at standard conditions (1 atmosphere, 70°F) are 1400 and 5580 scfm. The mass flowrates are:

$$\text{Air: 8000 acfm} \times 0.052 \text{ lb/ft}^3 = 420 \text{ lb/min}$$

$$\text{Water: 2000 acfm} \times 0.032 \text{ lb/ft}^3 = 65 \text{ lb/min}$$

(These densities were calculated at 1 atm. and 300°F.)

Thus, the inlet humidity is: $65/420 = 0.155$ lb/lb dry air. Using a psychometric chart, we begin at this humidity and 300°F and assume that the gas will be adiabatically cooled until it reaches saturation. In this case, we find that the gas reaches saturation at 148°F at a saturation humidity of 0.20 lb/lb. Because the air mass remains constant, the final water mass is $0.20 \times 420 = 84$ lb/min. The corresponding water volume is:

$$84 \text{ lb/min} \times 24.66 \text{ ft}^3/\text{lb} = 2070 \text{ acfm}$$

The dry air volume at 148°F is $420 \times 15.30 = 6430$ acfm.

The total flowrate is $2070 + 6430 = 8500$ acfm.

Now:

$$\text{Makeup} = \text{Bleed} + \text{Evaporation}$$

$$\text{Evaporation} = \text{Outlet} - \text{Inlet} = 84 - 65 = 19 \text{ lb/min} = 2.3 \text{ gpm}$$

From Equation 5.14:

$$\text{Bleed} = 0.120 \times 3.0 \text{ gr/dscf} \times 5580 \text{ dscf/min} \times 0.95 \times$$
$$1 \text{ lb/7000 gr} \times 4 \text{ lb water/lb solids} = 1.1 \text{ gpm}$$

$$\text{Makeup rate} = 2.3 + 1.1 = 3.4 \text{ gpm}$$

At a (relatively low) L/G of 5 gal/1000 acf, the recirculation rate would be $5 \times 8.5 = 42.5$ gpm. The above bleed rate would be less than 10% of the recirculation rate in that case.

Gaseous Scrubbers

Scrubbers are normally used to remove (absorb) gases of low concentration (up to about 3 volume percent) in the scrubbing liquid. Although spray towers, venturis, and other designs may be used, they are not as effective as packed or tray-type columns. The latter offer more intimate liquid-scrubber volume than the others. The dimensions required to size the column are the diameter, height, and wall thickness. The diameter (D_c, ft) is computed from the following:

$$D_c = 1.13(Q/v_r)^{0.5} \qquad (5.15)$$

where v_r = removal velocity (ft/min)
         Q = volumetric flowrate through column (acfm)

The removal velocity, determined empirically, must be low enough to prevent "flooding"—the condition when the gas flowing upward through the scrubber entrains the downward-flowing liquid. The velocities given in Table 5.9 take this into consideration.

The column height (H_c) is comprised of the removal section height (packing and supports or trays) and the added height needed for vapor/liquid separation, access for maintenance and inspections, etc. Thus:

$$H_c = H_r + H_a \qquad (5.16)$$

where H_r, H_a = removal and added heights (ft)

Typically, H_a = 2 to 3 ft plus 25% of the column diameter. However:

$$H_r = NTU \times HTU \qquad (5.17)$$

where   NTU = number of transfer units
         HTU = height of each transfer unit (ft)

For tray columns, NTU would be the number of trays, usually determined by a graphical method, such as McCabe-Thiele, or a numerical algorithm. The HTU would be just the tray spacing—typically, 1 to 2 ft.

For packed columns, closed-form expressions have been derived for calculating HTU and NTU. The HTU expression considers such variables as the gas and liquid flow rates, packing specifications, and the gas density, viscosity, and diffusivity. The NTU contains efficiency-related variables, such as Henry's constant, molar flow rates, and inlet/outlet concentrations of solute in the gas and liquid. These expressions can be found in reference 22 and elsewhere. Finally, the column thickness (T_c) is a complex function of the internal pressure, wind loading, and corrosion. Though it can range from 1/4 to 1-1/2 in., it typically runs from 1/4 to 1/2 in.

Most columns are costed on the basis of their total ("shell") weight (W_c, lb). Accordingly, we offer the following:

$$W_c = \pi D_c (H_c + 0.8116 D_c) T_c p_{pl} \qquad (5.18)$$

where $\qquad p_{pl}$ = density of material of construction (lb/ft^3)

The foregoing discussion merely "scratches the surface" of column sizing. In reality, column design is so complex that nearly all towers are custom-fabricated, making it extremely difficult to postulate a general sizing/costing procedure. Fortunately, vendors often offer small, packaged absorbers, each of which includes the column, internals, pumps, piping, etc. These units are usually sized on a gas flowrate (acfm) basis.

The liquid requirement for gas absorbers can be calculated from a material balance around the column. For the general case:

$$M_w(x_t - x_b) = M_g(y_b - y_t) \qquad (5.19)$$

where $\quad M_w$, M_g = water and gas mass flows, respectively (lb/min-ft^2 of column cross section)

x, y = mass fractions of solute (pollutant) in liquid and gas, respectively

t, b = subscripts denoting column top and bottom, respectively

(Recall here that the liquid and gas enter at the top and bottom of the column, respectively.)

From Equation 5.19 it is clear that the liquid/gas ratio (either mass or volume basis) will depend on the inlet and outlet concentrations. The solute-rich liquid is pumped from the column to storage or to another column where the solute is "stripped" from it. The stripped gas stream is much more concentrated than the inlet waste gas and, hence, is more economical to treat. Finally, the cleaned liquid is either recycled to the absorber or given final treatment (as needed) and discarded.

The horsepower requirement for an absorber is the sum of the system fan and pump(s) consumptions. In figuring the former, use the pressure drop that corresponds to the type of trays or packing being used. For example, Schifftner and Hesketh present pressure drops for about 30 different sizes and types of packings.[23] The pressure drops range from 0.13 to 1.60 in. w.c./ft of packing. Clearly, unless the packed height is significant, the gas pressure drop will not be enormous.

To estimate the pump(s) horsepower, assume a head of 100 feet per pump and figure the power from the following:

$$\text{Pump hp} = 0.000253 lh/n* \qquad (5.20)$$

where   l = liquid flowrate (gpm)
                 h = pump liquid head (ft of water)
                 n^* = pump efficiency (fractional)

(*Note:* Do not confuse the pump efficiency, "n^*," with the fan efficiency, "n'," or the combined fan-motor efficiency, "n." These last two terms are defined in Chapters 2 and 3, respectively.)

Costing Procedure

Equipment Costs

Most packaged scrubbers are priced according to the maximum gas volumetric flowrate they can accommodate. (In other words, the designs of packaged units incorporate the optimal removal velocities, L/G ratios, and other parameters discussed in the preceding section.) These prices include not only the scrubbers, but such auxiliaries as pumps, internal piping, and separators. We obtained vendor quotes for five types of scrubbers: venturis, impingement, horizontal cross-flow, vertical packed columns, and a "double venturi." In addition, a vendor supplied costs for mesh pad and Chevron mist eliminators. These costs are shown below. Where appropriate, we've also provided cost multipliers that allow us to estimate costs for units fabricated of different materials.

Venturis. Two vendors supplied quotes for venturis.[24,25] From these data, two price correlations were developed, each spanning a different flowrate range. The first correlation:

$$P(\$) = 8180 + 1.41Q \qquad (5.21)$$

where $600 \le Q, \text{acfm} \le 19{,}000$

And the second:

$$P(\$) = 84.2Q^{0.612} \qquad (5.22)$$

where $19{,}000 \le Q \le 59{,}000$

Each price (in **June 1988** dollars) is for an entire packaged scrubber system, including a venturi, mist eliminator, fan (direct or belt-driven), recirculation liquid pump, and sump. The material of construction is carbon steel.

Multiply these costs by the following (approximate) factors to obtain prices for units fabricated of other materials:

- Rubber lining: 1.6
- Fiber-reinforced plastic (FRP): 1.6
- Epoxy coating: 1.1

Notice that the first correlation is linear, while the second is a power function. This reflects the economies of scale that pertain to the sizing and pricing of larger scrubbers.

Impingement scrubbers. Prices were obtained from a vendor and escalated to **June 1988** dollars (from a February 1988 base).[26] Each price included a scrubber (3/16-in. carbon steel) with one, two, or three stages; 304 stainless steel baffle plates; internal sprays and piping; a fixed-blade mist eliminator; inspection doors; and inlet and outlet flanges. Costs for the fan, pumps, and other auxiliaries were NOT included. The prices fit equations of the form:

$$P(\$) = aQ^b \qquad (5.23)$$

where Q = maximum gas flowrate (acfm) such that: $885 \leq Q$
$\leq 76,950$
a, b = regression correlation parameters

Values for these parameters are:

Stages (#)	Effective Packing Height (ft)	a	b
1	8	53.4	0.570
2	16	62.4	0.586
3	24	63.0	0.610

This particular scrubber (a modified sieve plate) is designed to remove both gases and larger particles (1 μm and larger) quite efficiently. The pressure drop and recirculation requirements are 1.5 in. w.c. per stage and 2 to 3 gal/1000 acf, respectively. (The L/G values compare well with those in Table 5.10.) Finally, the maximum allowable solids content in the recirculating liquid is 10%.

To price impingement scrubbers with other materials of construction, multiply the above costs by the following factors:

- Coated carbon steel: 1.2 to 1.3
- FRP or PVC: 1.8 to 2.2

Horizontal cross-flow scrubbers. We also obtained costs for a cross-flow "fume washer" constructed of "corrosion-resistant plastics" (PVC, FRP, polypropylene, or combinations thereof).[27] Mounted on a plastic-coated steel base, each unit consists of a spray section followed by a 1-ft packed bed and, lastly, an impingement mist eliminator. Design specifications are: pressure drop—2.0 in. w.c.; removal velocity—500 ft/min; makeup water rate—0.5 gal/1000 acfm.

Escalated from December 1987 to **June 1988** dollars, the cross-flow scrubber prices fit the following expression:

$$P(\$) = 12.2Q^{0.716} \qquad (5.24)$$

where $800 \leq Q$, acfm $\leq 80,000$

This price includes the scrubber and all internals—packing, piping, mist eliminator, etc. Each unit is designed for remote locations and can accommodate subfreezing temperatures. However, the cost of a pump, piping, and plastic recirculation tank must be added to the base price. The pump, designed to overcome a 20-ft system head, is a corrosion-resistant, seal-less unit, with a magnetic drive. The tank volume required is a function of the recirculation rate and liquid residence time. The recirculation rate, in turn, depends on the gas flowrate and inlet/outlet solute concentrations. (See Equation 5.19.) Prices for recirculation units (in **June 1988** dollars) are:

$$P(\$) = 388V^{0.411} \tag{5.25}$$

where $\qquad$ V = tank volume (gal)
and $30 \leq V \leq 800$

Vertical packed columns. To accompany the above horizontal packed scrubber costs, we obtained prices for small vertical packed columns.[28] Each of these off-the-shelf columns is fabricated of FRP and contains 6 ft of polypropylene packing, a spray nozzle, liquid distributor, and a polypropylene mist eliminator. (No liquid storage equipment is included, as the vendor assumes that the liquid would not be recirculated.) The prices (in **September 1988** dollars) were correlated with the column diameter, D_c (ft) as follows:

$$P(\$) = 810 + 2180D_c \tag{5.26}$$

where $\qquad$ $1.0 \leq D_c \leq 2.5$

At an average removal velocity of 4 ft/sec. (see Table 5.9), these columns could treat from 200 to 1200 acfm of waste gas.

"Double venturi" scrubbers. Calvert Environmental Equipment[24] provided prices for their Collision Scrubber™. This scrubber consists of two adjustable venturi throats firing into each other, causing gases and liquid droplets to collide head-on. The action also causes the fine atomization of the droplets, since the liquid actually moves by inertia into the incoming gas stream.

Over a wide flowrate range, the following equation predicts the price (in **October 1988** dollars) for carbon steel Collision Scrubbers™:

$$P(\$) = 492Q^{0.450} \tag{5.27}$$

where $\qquad$ $2000 \leq Q, \text{acfm} \leq 120,000$

The prices of stainless steel scrubbers would be approximately twice the costs this equation provides. Included in these costs are the entrance duct, the Collision Scrubber™ throat, the diffuser, the entrainment separator, and all internals. Lastly, Collision Scrubbers™ operate with pressure drops of 25 to 50 in. w.c., and can recirculate slurries containing up to 10% solids.

Mist eliminators. A vendor supplied prices for several sizes of mesh pad and Chevron mist eliminators.[29] All costs are in **September 1988** dollars. However, the costs of the vessels in which the mist eliminators are installed are NOT included.

Mesh pad: Equation 5.28 predicts costs for a commonly used mesh pad eliminator. This unit has a 6-in.-thick 304 stainless steel bed that can capture 99% of particles larger than 7 μm. The removal velocity is 8 to 10 ft/sec at a pressure drop of 1 in. w.c.

$$P(\$) = 78.4D^{1.66} \tag{5.28}$$

where   D = mesh pad diameter (ft) and $2 \leq D \leq 10$

These prices include 1-in. support grids at the top and bottom of the mesh pad. For a given diameter, the price of the pad is roughly proportional to its thickness. For instance, the price of a 12-in.-thick × 6-ft-diameter pad would be $2670, versus $1540 for a 6-in. × 6-ft pad.

Chevron: Chevron eliminators are usually categorized according to style number, which in turn signifies the particle size the eliminator can remove at 99% efficiency. For example, a "Style 2" can remove particles >10 to 15 μm; "Style 1," $>20\,\mu$m; and "Style 8," $>40\,\mu$m. The following equation gives prices for "Style 2" Chevrons fabricated of 304 stainless steel:

$$P(\$) = 427 + 376D \tag{5.29}$$

where   D = Chevron diameter (ft) such that $0.5 \leq D \leq 3$

Auxiliary equipment (pumps). Some packaged scrubbers do not come equipped with recirculation pumps. Other times, an auxiliary pump may be needed to convey spent scrubber liquid to a storage, treatment, or disposal facility. In these (and similar) cases, pump prices would be needed to complete the control system capital cost estimate.

Accordingly, we obtained prices for a range of single-stage centrifugal pump sizes.[30] Each pump is cast iron/bronze fitted, contains a stainless steel mechanical seal, and includes either a single- or three-phase motor. These prices (escalated from November 1987 to **September 1988** dollars) were regressed against the motor horsepower (hp) to yield:

$$P(\$) = 538(hp)^{0.438} \tag{5.30}$$

where $0.5 \leq hp \leq 20$

Example: For the gas stream presented in the previous example, which scrubber would have the lower first (equipment) cost, a venturi or a "double venturi" (Collision Scrubber™)? Also, if the scrubber bleed water were to be disposed of in an onsite treatment pond, what size pump would be needed and what would it cost?

Solution: First, recall that the maximum gas flowrate through the scrubber is 8500 acfm and that the water bleed rate is 1.1 gpm. Upon substituting this gas flowrate into Equations 5.21 and 5.27, we obtain the following equipment costs (uninstalled):

$$\text{Venturi: } 8180 + 1.41(8,500) = \$20,200$$

$$\text{Collision: } 492(8500)^{0.450} = \$28,900$$

For the bleed water pump, assume a 100-ft head and a 60% pump efficiency. Substituting these values and the above bleed rate into Equation 5.20, we obtain a power requirement of only 0.04 horsepower. Therefore, we would have to use the smallest pump-and-motor size available (0.5.hp), the cost of which would be (according to Equation 5.30) $400.

From this simple example we see that (1) although the pressure drop (and attendant power consumption) of the Collision Scrubber™ is lower than the venturi's, its equipment cost is nearly 50% higher and (2) the power consumption and cost of the bleed water pump is negligible when compared to that of the scrubbers.

Total Capital Investment

As with fabric filters and electrostatic precipitators, the TCI of a wet scrubber system would be the product of the composite installation factor and the purchased equipment cost. However, as Table 2.2 shows, *two* sets of installation factors could apply to scrubbers: the "venturi" and the "gas absorber." The composite factor for venturis is 1.91 (1 + 0.56 + 0.35); the factor for gas absorbers, 2.20 (1 + 0.85 + 0.35). Which one should be used? In reality, either one could apply. And because the factors differ by only 15% or so, it probably wouldn't make much difference which were used. However, as a rough guideline, we'd suggest using the "venturi" installation factor for particulate scrubbers, and the "gas absorber" factor for gaseous scrubbers. This (somewhat arbitrary) division also applies to some of the direct annual costs listed below.

Annual Costs

Direct annual costs include expenditures for operating and supervisory labor, maintenance labor and materials, electricity, chemicals (e.g., caustic for ph control), water, and wastewater treatment. Again, unit prices for these commodities will vary, but consumption estimates can be given:

- Operating and maintenance labor (hr/shift), from Table 2.5:

Labor Category	Venturi	Gas Absorber
Operating	2 to 8	1/2
Maintenance	1 to 2	1/2

- Supervisory labor: 15% of operating labor
- Maintenance materials: 100% of maintenance labor
- Electricity: estimate consumptions for system fan, pumps, etc., from above horsepower equations and factor in a 90% motor efficiency.
- Chemicals: estimate requirement as needed. However, be sure to figure on basis of total water recirculation rate, not just makeup water rate.
- Wastewater treatment: The effluent from most scrubbers is usually too low to justify the construction of dedicated wastewater treatment systems. Consequently, scrubber wastewater treatment costs are usually assessed on a *pro rata* (e.g., $/1000 gal) charge basis. The underlying assumption here is that the onsite wastewater treatment system would have enough excess capacity to accommodate the scrubber wastewater.

A typical treatment system might consist of a gravity thickener, anaerobic digester, chemical conditioning unit, and vacuum filter. As the wastewater passes through this system, the solids concentration increases, so that by the time it exits the vacuum filter the effluent is a sludge with a 30 to 50 weight percent solids content. This sludge normally would be trucked to a landfill or land treatment facility for ultimate disposal. The capital and O&M costs of such a system would depend primarily on the wastewater flowrate and secondarily on the inlet wastewater suspended solids content. Hence, the *pro rata* charge would vary. Using cost data in reference 31, we obtained a wastewater system cost ranging from $0.01 to $0.08/1000 gal of wastewater influent.

This cost does not include the expense of trucking the dewatered sludge to a disposal site. This cost varies with the volume of sludge trucked and the round-trip distance to the site. For a typical hauling arrangement, the trucking cost would run from $0.05 to $0.20/yd^3-mi.[32] (*Note:* Both the treatment and trucking costs have been escalated to **June 1988** dollars.)

Procedures for calculating the *indirect annual costs* are the same as those presented in the "Fabric Filters" section. The only differences are (1) the capital recovery cost is calculated using the *entire* TCI, as no replacement parts costs apply, and (2) no recovery credits are included.

REFERENCES

1. *Chemical Engineering*, September 26, 1988, p. 34.
2. Turner, J. H., A. S. Viner, R. E. Jenkins, and W. M. Vatavuk. "Sizing and Costing of Fabric Filters, Part I: Sizing Considerations," *Journal of the Air Pollution Control Association*, June 1987, pp. 749–759 (hereinafter cited as "Baghouses, Part I").
3. "Baghouses, Part I."

4. "Baghouses, Part I."

5. "Baghouses, Part I."

6. Turner, J. H., A. S Viner, R. E. Jenkins, and W. M. Vatavuk. "Sizing and Costing of Fabric Filters, Part II: Costing Considerations," *Journal of the Air Pollution Control Association*, September 1987, pp. 1105-1112 (hereinafter cited as "Baghouses, Part II").

7. "Baghouses, Part II."

8. Turner, J. H., P. A. Lawless, T. Yamamoto, D. W. Coy, G. P. Greiner, J. D. McKenna, and W. M. Vatavuk. "Sizing and Costing of Electrostatic Precipitators, Part I: Sizing Considerations," *Journal of the Air Pollution Control Association*, April 1988, pp. 458-471 (hereinafter cited as "Electrostatic Precipitators, Part I").

9. "Electrostatic Precipitators, Part I."

10. "Electrostatic Precipitators, Part I."

11. "Electrostatic Precipitators, Part I."

12. Vatavuk, W. M., and L. Theodore. "A Comprehensive Technique for Calculating Particulate Control Device Efficiencies Utilizing Particle Size Distributions," *Proceedings of the Second National Conference on Energy and the Environment*, 1974, pp. 181-189.

13. Turner, J. H., P. A. Lawless, T. Yamamoto, D. W. Coy, G. P. Greiner, J. D. McKenna, and W. M. Vatavuk. "Sizing and Costing of Electrostatic Precipitators, Part II: Costing Considerations," *Journal of Air Pollution Control Association*, May 1988, pp. 715-726 (hereinafter cited as "Electrostatic Precipitators, Part II").

14. "Electrostatic Precipitators, Part II."

15. "Electrostatic Precipitators, Part II."

16. Schifftner, K. C., and H. E. Hesketh. *Wet Scrubbers: A Practical Handbook*. Chelsea, MI: Lewis Publishers, 1986 (hereinafter cited as *Wet Scrubbers*), p. 4.

17. *Wet Scrubbers*, p. 5.

18. *Wet Scrubbers*, pp. 30-44.

19. *Wet Scrubbers*, pp. 50-54.

20. *Wet Scrubbers*, pp. 27-28.

21. *Scrubber Selection Guide*. San Diego: Calvert Environmental Equipment, 1988.

22. Calvert, S., and H. M. Englund. *Handbook of Air Pollution Technology*. New York: John Wiley and Sons, 1984, pp. 366-368.

23. *Wet Scrubbers*, p. 45.

24. Price data from Calvert Environmental Equipment (San Diego, CA), October 1988.

25. Price and technical data from Intertel Corporation (Englewood, CO), August 1988.

26. Price and technical data from W.W. Sly Co. (Cleveland, OH), February 1988.

27. Price and technical data from Vanaire, Ltd. (Louisville, KY), December 1987.

28. Price and technical data from Croll-Reynolds, Inc. (Westfield, NJ), September 1988.

29. Price and technical data from Koch Engineering (Wichita, KS), September 1988.

30. Price and technical data from Multi-Duti Manufacturing, Inc. (Baldwin Park, CA), December 1987.

31. *Handbook: Estimating Sludge Management Costs*. Cincinnati: U.S. Environmental Protection Agency, October 1985 (EPA/625/6-85/010, NTIS PB-86-124542) (hereinafter cited as *Sludge Management Costs*), pp. 42-69, 81-83.

32. *Sludge Management Costs*, pp. 148-149.

CHAPTER 6

"Add-on" Controls III: Gaseous Control Devices

I counted two-and-seventy stenches
All well-defined, and several stinks. . .
Cologne—Samuel T. Coleridge

Where Chapter 5 dealt with particulate control devices, this chapter will cover *gaseous* control devices. As we've seen, all particulate controls collect pollutants. However, while some gaseous control devices are collectors, others are *converters*. "Collectors" include *gas absorbers*, *refrigerated condensers*, and *carbon adsorbers*. Similarly, *incinerators* (thermal, catalytic, and regenerative) and *flares* are pollutant "converters." Except for gas absorbers (covered in Chapter 5) this chapter will address the sizing and costing of these commonly used gaseous control devices.

INCINERATORS

Description

Incineration is a technique wherein a liquid, solid, or gas is oxidized (combusted) at high temperatures to form combustion products—primarily carbon dioxide (CO_2) and water (H_2O). When applied to the control of air emissions—mainly volatile organic compounds (VOCs)—incineration can be very effective, converting nearly 100% of combustibles present.

Fundamental Considerations

Broadly speaking, incineration is governed by the "three T's": *temperature*, *time*, and *turbulence*.[1] To achieve effective combustion and heat release rates without catalysts, the temperature of a combustible organic must be raised to 100°F or more above its ignition temperature (usually 1000 to 1400°F) and be held at this temperature for 0.3 to 1.0 sec. Further, turbulent conditions must be maintained in the incineration chamber to provide adequate mixing between the organics and the air used to oxidize them.

To combust most waste gases, *auxiliary heat* (in the form of oil, or preferably natural gas) must be provided, because the waste gas *heat content* frequently is too low to support combustion. The heat content, along with the

Table 6.1   Heat Contents of Selected Hydrocarbons at Their Lower Explosive Limits

Hydrocarbon	LEL (Vol. %)	Net Heat Content (Btu/scf)[a]
Alkanes		
Methane	5.0	44.3
Ethane	3.0	47.6
Propane	2.1	47.7
n-Butane	1.8	53.2
n-Pentane	1.4	51.0
n-Hexane	1.2	51.9
n-Heptane	1.05	52.6
n-Octane	0.95	54.0
n-Nonane	0.85	54.1
n-Decane	0.75	52.9
Alkenes		
Ethylene	2.7	39.7
Propylene	2.4	51.4
Butene-1	1.7	48.0
cis-Butene-2	1.8	50.7
Isobutylene	1.8	50.6
3-Methyl-butene-1	1.5	52.5
Aromatics		
Benzene	1.3	45.8
Toluene	1.2	50.3
Ethylbenzene	1.0	48.8
o-Xylene	1.1	53.5
m-Xylene	1.1	53.5
p-Xylene	1.1	53.5
Cumene	0.88	48.9
Cyclopentane	1.5	51.7
Cyclohexane	1.3	53.3
Methylcyclohexane	1.1	52.5
Ethylcyclohexane	0.95	51.9

Sources: 1. Perry, R. H., C. H. Chilton, and S. D. Kilpatrick, Eds. *Chemical Engineers' Handbook*, 4th ed. New York: McGraw-Hill, 1963, pp. 3–142 to 3–144. 2. *Handbook of Chemistry and Physics*, 54th ed. Boca Raton, FL: CRC Press, Inc., 1973–74, pp. D85-D92.
[a]Expressed as Btu/scf of *mixture* (hydrocarbon and air) at lower explosive limit (LEL).

combustion temperature, fuel heating value, and other parameters, determines the auxiliary heat requirement. For a given waste gas, the heat content is primarily a function of the gas composition and oxygen content. The last two parameters establish the "flammability characteristics" of the waste gas, which is usually expressed by the "lower explosive limit" (LEL) and the "upper explosive limit" (UEL). These limits represent, in turn, ". . .the smallest and largest amounts of VOCs which, when mixed with air, will burn without a continuous application of heat."[1]

Moreover, for most *hydrocarbons*, the heat content at the LEL is approximately 50 Btu/scf of mixture. For instance, at 25% of the LEL—the maximum allowable inlet concentration for incinerators—the heat content would be approximately 13 Btu/scf. Table 6.1 lists the heat of combustion for selected hydrocarbons measured at their respective LELs.[2,3] Note that the heats are close to the 50-Btu/scf rule of thumb, varying from it by less than 10% in most cases.

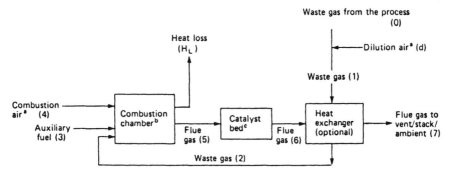

(a) When required.

(b) Referred to as preheat chamber in
the case of catalytic incinerators.

(c) Included only in catalytic incinerators.

Figure 6.1   Diagram of a thermal/catalytic incineration system. (Numbers indicate locations in system.) (Reprinted with permission from *Journal of the Air Pollution Control Association*, January 1987, pp. 91–99.).

If the waste gas concentration should exceed 25% LEL, dilution air would have to be added before the waste gas enters the incineration system. Thus, the 25% LEL limit constrains an incinerator's operation. So does the waste gas inlet oxygen content. A waste gas with less than 13–16% oxygen requires auxiliary air just to maintain burner stability, let alone to combust the organics present.

Incineration Equipment

The most commonly used air pollution control incinerators are *thermal* and *catalytic*. Another type used, the *regenerative thermal oxidizer* (RTO), is a variant of the thermal design.

Both thermal and catalytic incinerators employ a similar combustion process, design procedure, and equipment. One major difference: in catalytic units, the combustion is essentially flameless, wherein the conversion takes place on a catalyst bed rather than in an empty, cylindrical combustion chamber. Prior to entering the incinerator combustion chamber (or "preheat" chamber, in catalytic units), the waste gas almost always passes through a *recuperative heat exchanger*, where it is preheated by exiting combustion gases. (See Figure 6.1 for a schematic of this system.) The preheat temperature (T_2 in Figure 6.1) is limited to 1000 to 1100°F, above which temperature preignition could occur. (*Note*: The subscripts for T and Q correspond to the locations in the incinerator system, as shown in Figure 6.1.) Preignition can cause severe physical damage to the heat exchanger, due to rapidly expanding gases. This heat exchanger serves to reduce the auxiliary heat requirement. A second heat recovery apparatus may be used downstream to recover additional heat. However, this "secondary" recovery is not feasible unless there is a use for the additional heat recovered either on- or offsite.

As stated above, combustion temperatures should exceed the ignition temperatures by at least 100°F. For thermal incinerators, this translates to combustion temperatures on the order of 1400 to 1600°F, at which 98 to 99% of VOCs are combusted, respectively. Corresponding residence times of 0.50 to 1 second are recommended. However, due to the presence of catalyst, the conversion temperature in catalytic incinerators is much lower: 700 to 900°F (But in no case can the catalyst bed temperature exceed 1200°F.) The corresponding preheat temperature is usually 500 to 600°F.[1]

The catalyst itself is either a precious metal or metal salt (platinum, palladium, rhodium, nickel, or gold) or a base metal (primarily manganese dioxide). As installed, the catalyst is either supported on an inert carrier or unsupported. Because they gradually lose their effectiveness ("activity"), catalysts must be replaced after several years. Further, catalysts are subject to poisoning by phosphorous, arsenic, etc., and plugging by particulate matter. This limits the application of catalytic incinerators.

Regenerative thermal oxidizers, such as those manufactured by Salem Industries, Inc., are different from catalytic and thermal incinerators in terms of both design and operation, although the fundamental combustion principles governing them are the same. In these incinerators, the entering waste gas passes vertically through a chamber filled with several feet of ceramic or stoneware packing (bed #1) wherein it is warmed by several hundred degrees Fahrenheit. The residence time in this chamber is at least 0.5 sec. Some of the VOC burns in this preheat chamber, while the rest is burned in a central combustion chamber, at a temperature of 1400 to 2200°F. Following combustion, the waste gas exits through another packed chamber (bed #2) to which it contributes most of its enthalpy. The beds are then switched, so that the next increment of waste gas enters through bed #2 for preheating. Thus, each bed is cooled, heated, cooled again, etc. Each cooling-heating cycle lasts a few seconds, so that each bed—and the entire RTO itself—approaches steady-state conditions. Although the capital cost of regenerative oxidizers is higher than the thermal and catalytic incinerator types, the heat recuperation efficiency is much higher: up to 95% vs. a maximum of 70% in traditional incinerators.[4] (A typical RTO is shown in Figure 6.2.)

Sizing Procedure

Primary Sizing Parameters

The *flue gas flowrate* and *auxiliary fuel requirement* are the most important sizing parameters for a thermal or a catalytic incinerator. The former determines the equipment size and cost, while the latter comprises most of the annual operating and maintenance cost. These parameters are interdependent, in the sense that their calculation is based on material and energy balances

Figure 6.2 Regenerative thermal oxidizer (courtesy Salem Industries, Inc.).

taken around the incinerator. Reference 1 presents these balances in exquisite detail, for each of six waste gas "categories." These "categories" are delineated according to the waste gas oxygen content, VOC concentration, and heat content. For "category 1" waste gas streams—those most commonly encountered in air pollution control applications—the flue gas flowrate (Q_5, scfm) and auxiliary heat requirement (Q_3, scfm) are calculated by the next two equations:

$$Q_5 = Q_2 + Q_3 \tag{6.1}$$

where Q_2 = inlet waste gas flowrate (scfm)

(Category 1 streams consist of a mixture of VOC, air, and inert gas, in which the oxygen content exceeds 16% *and* the VOC content is <25% LEL. Hence, neither dilution nor auxiliary combustion air is needed.)

$$Q_3/Q_2 = X/Y \tag{6.2}$$

where

$$X = 1.1C_{p5}(T_5 - T_r) - C_{p2}(T_2 - T_r) - h_1 \tag{6.3}$$
$$Y = h_3 - 1.1C_{p5}(T_5 - T_r) \tag{6.4}$$

C_{p5}, C_{p2} = mean heat capacities of streams leaving and entering combustion chamber, respectively (Btu/scf-°F)

T_5, T_2 = combustion chamber, waste gas inlet temperatures, respectively (°F)

(*Note:* For catalytic incinerators, use T_6 in place of T_5.)

T_r = reference (= fuel inlet) temperature (typically, 70°F)

h_1, h_3 = waste gas heat content and fuel *net* heating value, respectively (Btu/scf)

Note that Equation 6.2 is a ratio, with units of standard cubic feet of fuel per standard cubic foot of waste gas. Also, two heat capacities are given, because C_ps vary with temperature. Example values for C_p are:

Temperature (°F)	Mean Heat Capacity of Air (Btu/scf-°F)
100	0.018
1600	0.0194
1800	0.0196
2000	0.0198

(*Note:* The heat capacity of air is used because air comprises well over 90% of the waste gas, before and after combustion.)

Finally, T_2 is calculated from either of these expressions:

$$\text{Thermal: } T_2 = T_1 + \&(T_5 - T_1) \tag{6.5}$$

$$\text{Catalytic: } T_2 = T_1 + \&(T_6 - T_1) \tag{6.6}$$

where

T_1 = waste gas temperature *entering* heat exchanger (°F)

$\&$ = heat exchanger efficiency (fractional)

Vendors can custom-build heat exchangers to achieve nearly any heat recuperation efficiency. Nonetheless, most modular heat exchangers have one- to three-pass configurations, with these efficiencies:[1]

No. of Passes	Efficiency, & (%)
1	35 to 40
2	45 to 50
3	65 to 70

What value of & should we select? Good question. First, as & increases, so does T_2, while the auxiliary fuel requirement (and fuel cost) decreases. But as & increases, the cost of the heat exchanger goes up, and, in turn, the system total capital investment (TCI). Thus, there is a trade-off between operating and capital cost, the extent of which depends on the T_2 selected. (Recall also that T_2 is limited to 1000 to 1100°F.) Such a selection needs to be based on a comprehensive process optimization, the details of which are beyond the scope of this book.

For regenerative thermal oxidizers (RTOs), the heat recuperation efficiency, $\&_{RTO}$ (or thermal efficiency, T.E.) is calculated as follows:[4]

$$\&_{RTO} = (T_5 - T_6)/(T_5 - T_1) \qquad (6.7)$$

where T_6 = outlet (flue gas) temperature (°F)

At first glance, this is different from Equation 6.5. However, an energy balance around the thermal incinerator recuperative heat exchanger (see Figure 6.1) would show that $T_5 - T_6 = T_2 - T_1$, so that Equations 6.5 and 6.7 are essentially equivalent.

As we said earlier, $\&_{RTO}$s are quite high, typically ranging from 85 to 95%. Hence, the RTO fuel requirement is usually quite low. It can be shown that the *net* fuel requirement (F_n, Btu/hr) is the *gross* fuel requirement (F_g, Btu/hr) minus the energy recovered minus the total heats of combustion of the VOC present in the inlet waste gas (h_1, Btu/hr). Hence:

$$F_n = 1.1(1 - \&)F_g - h_1 \qquad (6.8)$$

However:

$$F_g = Q_2 C_{p5}(T_5 - T_1) \qquad (6.9)$$

Upon combining Equations 6.7, 6.8, and 6.9, we obtain:

$$F_n = 1.1 Q_2 C_{p5}(T_6 - T_1) - h_1 \qquad (6.10)$$

where the "1.1" factor accounts for heat losses in the oxidizer.

To calculate the volume of natural gas needed, divide F_n by Y. (See Equation 6.4.)

Finally, due to the low fuel requirement, the RTO inlet and outlet gas flowrates are approximately equal. Hence, RTOs are usually sized on the basis of their *inlet* flowrates (Q_2).

Example: Given a 1400°F RTO combustion temperature (T_s), a 350°F inlet temperature (T_1), and a 95% thermal efficiency (&), calculate the net fuel requirement (F_n) for a 10,000-scfm waste gas stream containing 250 ppm of toluene. Assume a 10,000 Btu/lb heat of combustion for toluene.

Solution: First, calculate the toluene inlet mass rate:

Mass in = (250/1,000,000)(10,000)(60 m/hr)(0.238 lb/ft³)

   = 35.7 lb/hr

Waste gas heat content = (35.7 lb/hr)(10,000 Btu/lb)

   = 357,000 Btu/hr

Next, substitute the thermal efficiency (0.95), combustion temperature, and inlet temperature into Equation 6.7 and solve for T_6:

$$T_6 = 403°F$$

Finally, via linear interpolation, we calculated the mean heat capacity (C_{ps}) between 70 and 1400°F to be 0.0192 Btu/scf-°F. Substituting this value and the other variables into Equation 6.10, we obtain:

$$F_n = (1.1)(10,000)(0.0192)(403 - 350)(60) - 357,000$$

   = 315,000 Btu/hr

Assuming a net heating value (h_3) of 900 Btu/scf, what volume of natural gas would be required to provide this much energy? To find out, just divide F_n by Y, where:

$$Y = h_3 - 1.1C_{ps}(T_5 - T_r)$$
$$= 900 - (1.1)(0.0192)(1400 - 370)$$
$$= 872 \text{ Btu/scf}$$

Thus, natural gas volume = 315,000/872 = 361 scf/hr.

Pressure Drop

The total pressure drop through an incinerator unit is the sum of the losses through the incinerator and heat exchanger. These losses depend on the gas flowrate, temperature, viscosity, and other variables. But for budget-study estimating purposes, the following pressure drops are sufficiently accurate:[1]

- Incinerators
  thermal: 4 in. w.c.
  catalytic: 6 in. w.c.
- Heat exchangers
  35%-efficient: 4 in. w.c.

50%-efficient: 8 in. w.c.
70%-efficient: 15 in. w.c.

For example, a thermal incinerator with a 50%-efficient heat exchanger would have a total pressure loss of 12 in. w.c. (4 + 8).

For regenerative thermal oxidizers, the pressure drop also depends on the thermal efficiency. The following pressure drops are representative:[4]

- 85% thermal efficiency: 16 in. w.c.
- 95% thermal efficiency: 20 in. w.c.

Catalyst Requirement

The amount of catalyst needed depends upon the catalyst type and age, required control efficiency, and potential contaminants present. For estimating purposes, these values may be used:[1]

Catalyst	Control Efficiency (%)	Amount Needed (ft³/1000 scfm)
Precious metal	90	1.5
	95	2
Base metal	90	4
	95	6

Costing Procedure

Equipment Costs

Thermal and catalytic incinerators. Most thermal and catalytic incinerators handling waste gas flowrates up to 30,000 scfm are usually sold as packaged units. Each of these packaged units consists of: (1) the combustion unit and burner (carbon steel or stainless steel refractory chamber for thermal units, preheat chamber and catalyst bed with stainless steel enclosure for catalytic units); (2) heat exchanger; (3) fan and motor (waste gas fan and, if needed, ambient air fan); (4) instrumentation and controls; (5) 10-ft stack; and (6) (catalytic units only) filter/mixer to distribute the gas flow, protect the catalyst bed from flame impingement, and remove noncombustible particulate matter. Ductwork and other auxiliary equipment "sold separately" may also be required.[5]

Costs for packaged thermal and catalytic incinerators were taken from reference 5 and escalated from June 1986 to **June 1988** dollars. These incinerators include all of the equipment listed above and are designed for burning dilute (<25% LEL) VOC streams at 1 atmosphere pressure and a 1500°F combustion temperature (thermal) or 600°F preheat temperature (catalytic). However, these costs would still apply to slightly different operating temperatures (say, a 1700°F combustion temperature).

The *thermal incinerator* prices (P) fit curves of the following form:

Table 6.2   Parameters for Thermal and Catalytic Incinerator Price Equations

Heat Exchanger Efficiency (%)	Regression Parameters	
	a	b
Thermal incinerators[a]		
0	3,120	0.360
35	4,760	0.375
50	4,920	0.389
70	5,690	0.408
Catalytic incinerators[b]		
0	11.3	3.57×10^{-5}
35	11.6	3.42×10^{-5}
50	11.7	3.54×10^{-5}
70	11.8	3.64×10^{-5}

Source: Katari, V. S., W. M. Vatavuk, and A. H. Wehe. "Incineration Techniques for Control of Volatile Organic Compound Emissions, Part II: Capital and Annual Operating Costs," *Journal of the Air Pollution Control Association*, February 1987 (Vol. 37, No. 2), pp. 198–201.
[a]Regression parameters fit equations of the form:
 Price ($) = aQ^b (Q = flue gas flowrate, scfm)
[b]Regression parameters fit equations of the form:
 Price ($) = exp(a + bQ)

$$P(\$) = aQ_5{}^b \qquad (6.11)$$

where Q_5 = flue gas flowrate (scfm)
          a, b = regression parameters whose values depend on the
               heat exchanger efficiency

The *catalytic incinerator* prices fit a different equational form:

$$P(\$) = \exp(a + bQ_5) \qquad (6.12)$$

Equations 6.11 and 6.12 are only valid for flowrates between 5000 and 50,000 scfm. Values for the parameters a and b are listed in Table 6.2.

Regenerative thermal oxidizers. Upon regressing cost quotes obtained from a RTO vendor,[4] we developed two cost correlations (in **December 1988** dollars) for units with 85% and 95% thermal efficiency (T.E.), respectively:

$$85\% \text{ T.E.: } C_{RTO}(\$) = 383{,}000 + 15.3Q_2 \qquad (6.13)$$

$$95\% \text{ T.E.: } C_{RTO}(\$) = 464{,}000 + 19.1Q_2 \qquad (6.14)$$

where Q_2 = *inlet* waste gas flowrate (scfm)
               such that $5000 \leq Q_2 \leq 500{,}000$
          $C_{RTO}(\$)$ = *installed* cost of RTO

Two features of these equations distinguish them from the other incinerator cost formulas. First, the equations estimate the *installed* costs of the RTOs — i.e., the costs of all equipment and labor needed to erect, assemble, and start up the unit onsite. The only costs not included therein are the costs of the foundation, ductwork, freight, and sales taxes. According to reference 4, the costs of ductwork, freight, etc., typically would add another 15 to 20% to the base cost of the RTO.

Also note that the costs have been correlated with the *inlet* waste gas flow-rate, rather than the flue gas flowrate. As explained above, that is because the inlet and flue gas flows are essentially equal for most streams burned in RTOs.

Total Capital Investment

Packaged thermal and catalytic incinerators designed to handle flows up to 30,000 scfm are usually modular, skid-mounted units that are easy and inexpensive to install at the site. In most cases, all the customer has to furnish is a foundation, the ductwork, and the labor and materials required to connect the incinerator to onsite utilities. For skid-mounted units, therefore, the TCI is a relatively low multiplier of the PE (purchased equipment cost). Reference 5 recommends a composite installation multiplier of 1.25 times the PE. (Recall from Chapter 2 that the PE is the sum of the incinerator and auxiliary equipment costs and the costs of instrumentation, sales taxes, and freight.)

Larger incinerators, handling flows up to 100,000 scfm, also have been built. But because these are "custom" units that are part vendor-fabricated and part field-assembled, their installation factors are higher. For these, the composite installation factor listed in Table 2.2 would apply. That is, TCI = 1.61PE. Finally, for regenerative thermal oxidizers, figure the TCI at 1.15 to 1.20 times C_{RTO}. As stated above, this factor would include the cost of ductwork, as well as the foundation, freight, taxes, and other installation costs.

Annual Costs

Thermal and catalytic incinerators: direct. *Direct* annual costs for thermal and catalytic incinerators include labor (operating, supervisory, and maintenance), maintenance materials, electricity, and last but not least, fuel. Also, for catalytic incinerators, the cost of replacement catalyst applies. Although the unit prices (e.g., $/kWh) will vary by installation, the following general consumption figures can be given:[5]

- Operating labor: 0.5 hr/shift
- Supervisory labor: 15% of operating labor
- Maintenance labor: 0.5 hr/shift
- Maintenance materials: 100% of maintenance labor
- Electricity: Compute fan power requirements from Equation 2.2, inputting the pressure drops listed above.
- Fuel: Calculate this from Equations 6.2, 6.3, and 6.4, above.
- Catalyst requirement: Because the catalyst is replaced less often than annually, its first cost should be treated as a separate investment with a lifetime different from that of the rest of the incinerator system. (In

this sense, the catalyst is like the bags in a fabric filter. See Chapter 5.) The life of a catalyst bed ranges from 2 to 10 years, with the life typically falling closer to the lower end of the scale.[5] Calculate the catalyst replacement cost (C_c) as follows:

$$C_c = (P_c + P_{cl})CRF_c \qquad (6.15)$$

where P_c = initial price of catalyst, including taxes and freight ($)

P_{cl} = labor cost to replace catalyst ($)

CRF_c = capital recovery factor for useful life of catalyst and interest rate in question. (See Chapter 2 for discussion of CRF.)

The initial costs of catalysts are $3000/ft^3 and $600/ft^3 of catalyst, respectively, for precious metal and base metal catalysts. Add to these prices the costs of freight and sales taxes to determine P_c. Estimate the catalyst requirement using the factors given in the "Sizing Procedure" above. Finally, the catalyst replacement labor cost, P_{cl}, is a relatively small fraction of the catalyst price — usually < 10%.

Regenerative thermal oxidizers: direct. RTOs incur the same kinds of direct annual costs as other incinerators. For operating and supervisory labor, use the same values as for thermal and catalytic incinerators. However, reference 4 suggests a lower value for *maintenance labor* — 1 hr/operating week. This maintenance labor includes such work as the daily oiling of fan bearings, the monthly and quarterly cleaning of valve seats, and the adjusting of burners. As above, assume the cost of maintenance materials to equal the maintenance labor cost. Finally, calculate the electricity cost using the RTO pressure drops given previously and (as shown in the above example) compute the fuel expenditure via Equations 6.7 through 6.10.

Indirect annual costs. Indirect annual costs for *all* incinerators include capital recovery, overhead, property taxes, insurance, and administrative charges. As Chapter 2 recommends, estimate the last three items at a total of 4% of the TCI and figure overhead at 60% of the sum of all labor and maintenance materials costs.

To calculate the capital recovery cost for thermal incinerator and RTO systems (CRC_s), multiply the TCI by the system capital recovery factor (CRF_s). The CRF_s is based on the annual interest rate and the system lifetime — 10 years for thermal and catalytic incinerators[5] and 20 years for RTOs.[4]

Use the same procedure for catalytic incinerators, except first adjust the TCI as shown below:

$$CRC_s = (TCI - P_c - P_{cl})CRF_s \qquad (6.16)$$

This adjustment is necessary to avoid "double counting" in the capital recovery calculation.

Finally, the *total annual cost* (TAC) is the sum of the direct and indirect annual costs. (There are no recovery credits or disposal costs to consider.)

Example: A coil coating spray booth emits 10,500 scfm (at T_1 = 100°F) containing a 50–50 mixture of toluene and *m*-xylene in air. The total hydrocarbon concentration is 0.27% (2700 ppm). The waste gas is to be burned in a packaged, natural gas–fired thermal incinerator with a recuperative heat exchanger. Estimate the capital and annual costs of the incinerator-heat exchanger, assuming a 1600°F combustion temperature (T_5), a duct length of 150 ft, and an 8000 hr/yr operating schedule. The facility is located in piedmont (central) North Carolina.

Solution:

1. *Waste stream categorization:* From Table 6.1, the heat contents at 100% LEL for toluene and *m*-xylene are 50.3 and 53.5 Btu/scf, respectively. The corresponding LELs are 1.2% (toluene) and 1.1% (*m*-xylene). And, because this is a 50–50 mixture, the concentration of each constituent is 0.135% (0.27/2). Thus, assuming a proportionality between %LEL and heat content for each hydrocarbon, the total heat content (h_1) of the waste gas stream would be:

$$h_1 = (0.135/1.2)50.3 + (0.135/1.1)53.5 = 12.2 \text{ Btu/scf}$$

Is this waste stream at or below the 25% LEL level? To be conservative, let's assume that the entire stream is *m*-xylene, because it has the lower LEL. Then:

$$\%LEL = 0.27/1.1 \times 100 = 24.5\%$$

Because the stream is below the 25% LEL cutoff, it can be burned as is. If the concentration were higher, however, dilution air would've had to be added.

2. *Flue gas flowrate and natural gas requirement:* The first step is to calculate the combustion chamber inlet temperature, T_2. But this temperature depends on the heat exchanger efficiency—a value that was not given. As a first approximation, assume maximum heat recovery (70%). Substituting into Equation 6.5:

$$T_2 = 100 + 0.70(1500 - 100) = 1150°F$$

Note that T_2 is approximately at the recommended maximum inlet temperature. Clearly, a higher heat recovery could not be tolerated. An energy balance around the heat exchanger would show that the temperature rise in the incoming waste gas stream must equal the temperature drop in the outgoing flue gas stream. Therefore, the flue gas outlet temperature (T_6) must be 1600 – 1050 = 550°F.

To calculate the natural gas requirement, substitute the values for T_2, T_1, and h_1 into Equation 6.2, along with the heat capacities C_{ps} (0.0194 at 1600°F) and C_{p2} (0.019 at 1150°F), the reference temperature (T_r = 70°F), and the natural gas NET heating value, h_3. For the last, assume 900 Btu/scf. Finally:

$$Q_3/Q_2 = -0.07/867.3 = -8.1 \times 10^{-5} \text{ scf natural gas/scf waste gas}$$

Thus, the natural gas requirement is essentially zero. However, a small amount of gas will be needed to stabilize the burner. This amount is approximately 5% of the total heat input ($C_{ps}[T_5 - T_r]$) or 1.5 Btu/scf of waste gas. This corresponds to 0.0017 scf/scf waste gas.

$$\text{And: } Q_3 = (0.0017)(10,500) = 18 \text{ scfm}$$

From Equation 6.1:

$$Q_5 = Q_2 + Q_3 = 10,518 \text{ scfm}$$

This flowrate, Q_5, will be used to size the incinerator.

3. *Equipment costs:* Because this is a packaged incinerator, all the required equipment is included in the vendor's price, except the ductwork. As Table 6.2 shows, the price equation parameters for a thermal incinerator with a 70%-efficient heat exchanger are: a = 5,690; b = 0.408. Substituting these parameters into Equation 6.11:

$$\text{Price (\$)} = 5,690(10,518)^{0.408} = \$248,900$$

Regarding the ductwork price, recall that the duct length given was 150 ft. The duct diameter would be based on the waste gas (not the flue gas) flowrate, in acfm (not scfm). At 100°F, this flowrate is: 10,500(560/530) = 11,100 acfm. At an inlet velocity of 2000 ft/min, the required duct diameter (D, in.) would be:

$$D = 13.54(11,100/2,000)^{0.5} = 31.9 \text{ in.} = 2.66 \text{ ft}$$

As Chapter 4 indicates, the price of carbon steel duct of diameters between 2 and 4 ft is $1.03/lb. The weight (W) for 1/4-in. carbon steel duct is provided by Equation 4.13. Combining all this yields:

$$\text{Duct price (\$)} = (\$1.03/\text{lb})(32.0)(2.66 \text{ ft})(150 \text{ ft})$$
$$= \$13,200$$

The total base equipment cost is: $248,900 + $13,200 = $262,100.

3. *Total capital investment:* Because this is a packaged incinerator, we can use the lower installation factor given above. This factor (1.25) is multiplied by the purchased equipment cost (PE), which, in turn, is the sum of the base equipment cost, sales taxes, and freight. (The incinerator

price includes instrumentation.) Recalling that sales taxes and freight total 8% of the base equipment cost, we have:

$$TCI = (1.25)(1.08)(\$262,100) = \$354,000 \text{ (rounded)}$$

4. *Total annual cost (TAC):* As this facility is located in piedmont North Carolina, we can obtain representative unit prices for labor, natural gas, etc. With these prices, along with the above general consumption figures and the operating factor (8000 hr/yr = 1000 shifts/yr), we can calculate the direct annual costs as follows:

- Operating labor = ($8/hr)(0.5 hr/sh.)(1000 sh./yr)
    = \$4,000.
- Supervisory labor = (0.15)($4,000) = \$600.
- Maintenance labor = ($8.80)(0.5)(1000) = \$4,400.

(*Note:* the maintenance labor rate is estimated at 110% of the operating labor rate.)

- Maintenance materials = maintenance labor = \$4,400.
- Electricity: Electricity is consumed by the system fan, located downstream of the unit. As indicated above, the total pressure loss (ΔP) is 4 in. (incinerator) + 15 in. (heat exchanger) + 0.4 in. (ductwork) = 19.4 in. (The ductwork pressure loss was calculated from Equation 4.14.) Assuming an overall (conservative) 50% fan-motor efficiency, the fan motor power requirement (M, kW) would be:

$$M \text{ (kW)} = (0.746 \text{ kW/hp})(0.0001575)(1/0.50)\Delta P Q_s$$

where Q_s = flue gas flowrate (acfm)
       = (10,518)(1010/530) = 20,040 acfm

Substituting the above values and a $0.05/kWhr electricity price:

Electricity cost = ($0.05/kWhr)(91.4 kW)(8,000 hr/yr)
       = \$36,600/yr

- Natural gas: In step 2 above, we calculated the natural gas requirement per unit volume of waste gas. Based on the above 900 Btu/scf net heating value and a gas price of $5.00/million Btu, we have:

Gas cost = ($5.00)(900)(0.0017)(10,500)(60)(8,000) = \$38,800/yr

The sum of the direct annual costs is \$88,800.

The indirect annual costs include capital recovery, property taxes, insurance, administrative charges, and overhead.

- Capital recovery: Assuming a 10-year system life and a 10% annual interest rate, we have a capital recovery factor (CRF) of 0.1628. The capital recovery cost (CRC) is the product of the CRF and the TCI, or:

$$CRC = (0.1628)(\$354,000) = \$57,600$$

- Property taxes, insurance, and administrative charges: Totalling 4% of the TCI, these costs are:

$$(0.04)(\$354,000) = \$14,200$$

- Overhead: This is computed as 60% of the sum of the costs for operating, supervisory, and maintenance labor and maintenance materials, or:

$$Overhead = (0.60)(\$4000 + \$600 + \$4400 + \$4400)$$
$$= \$8100/yr$$

The sum of the indirect annual costs is $79,900.

Finally, the total annual cost (rounded) is: $169,000.

Note that the natural gas cost ($38,000) accounts for only 23% of the TAC. This is somewhat atypical of thermal incinerators. In many such incinerators, the gas cost can contribute as much as 80% of the TAC. In this case, the high waste gas heat content, coupled with a high heat recovery (70%), resulted in a theoretical fuel requirement of essentially zero. The small amount of gas required for burner stabilization had a relatively weak impact on the TAC. In fact, the gas cost was so low that the direct and indirect annual costs were roughly equivalent—also atypical for thermal incinerators.

What if we had selected a catalytic incinerator instead of a thermal unit? Would its TAC have been lower? We'll let the reader answer that question.

CARBON ADSORBERS

Sometimes the gaseous emissions present in the waste gas are valuable enough to recover for recycling or resale. Other times, the waste gas volume may be too large to economically incinerate, due to the high cost of fuel. On such occasions, *collection* methods may be appropriate to use. This chapter next will cover two of the most commonly used collection devices—carbon adsorbers and refrigerated condensers. Adsorbers are suited to controlling streams of low to medium pollutant concentrations, while condensers are used to control higher concentration offgases—especially organic vapors.

Description

If gas molecules are passed through a bed of solid particles and some of them are collected on this bed, the captured molecules are said to be *adsorbed*. Adsorption occurs due to weak attractive ("van der Waals") forces, which are weaker (and less specific) than chemical bonds. As the molecules ("adsorbate") are at a lower energy state after capture than before, the process releases energy—the "heat of adsorption"—which is approximately equal to the heat of condensation. Several classes of organic compounds are readily "adsorbable," including hydrocarbons (halogenated and nonhalogenated), alcohols, esters, and ketones.[6]

Although several types of adsorbents are used, activated carbon is the most prevalent. Activated carbon is made by first burning petroleum or coal in a reducing atmosphere and then subjecting it to high pressure steam to create thousands of tiny, interconnecting capillaries in the material. The adsorbate molecules are captured in these labyrinthine capillaries, as the waste gas flows through the adsorbent bed.[6]

The amount of adsorbate captured on the carbon bed depends on several variables, including the inlet adsorbate concentration; operating temperature, pressure, and moisture content; characteristics of the adsorbent (e.g., bulk density); velocity of the offgas; and the molecular weight, boiling point, and chemical properties of the adsorbate. For a given set of operating conditions (temperature, pressure, etc.) and carbon type, the maximum (or equilibrium) amount captured is solely a function of the inlet concentration (usually expressed in parts-per-million volume—ppmv), which constitutes the adsorption driving force. For a range of inlet concentrations, we can measure the corresponding amounts adsorbed at equilibrium ("equilibrium adsorptivity"). The plot of these data form an *adsorption isotherm*. The mathematical forms of isotherms vary. For certain types of isotherms, a power function accurately predicts the equilibrium adsorptivity (m_e, lb adsorbate/lb adsorbent) over limited ranges of adsorbate partial pressure (i.e., inlet concentration):

$$m_e = ap^b \qquad (6.17)$$

where p = partial pressure of adsorbate in waste gas stream (psia)
a, b = isotherm parameters

The isotherm parameters vary according to the adsorbate, type of carbon, and adsorption temperature. As a rule, the higher the temperature, the *lower* the equilibrium adsorptivity. Table 6.3 lists the isotherm parameters for selected organic compounds. Note that these equations may NOT be extrapolated outside the partial pressure ranges shown. Adsorptivity data for other compounds, partial pressures, and temperatures are available from activated carbon vendors and such handbooks as reference 3.

Example: Estimate the equilibrium adsorptivity for *m*-xylene on Calgon "BPL" carbon (4 × 10 mesh) at two inlet concentrations (in air): 60 and 2000 ppmv. The adsorption temperature and pressure are 77°F and 1 atmosphere, respectively.

Table 6.3 Isotherm Parameters for Selected Organic Adsorbates[a]

Adsorbate	Adsorption Temperature (°F)	Isotherm Parameters		Partial Pressure Range (psia)[b]
		a	b	
Acetone	100	0.412	0.389	0.0001 – 0.05
Acrylonitrile	100	0.935	0.424	0.0001 – 0.015
Benzene	77	0.597	0.176	0.0001 – 0.05
Chlorobenzene	77	1.05	0.188	0.0001 – 0.01
Cyclohexane	100	0.508	0.210	0.0001 – 0.05
Dichloroethane	77	0.976	0.281	0.0001 – 0.04
Phenol	104	0.855	0.153	0.0001 – 0.03
Toluene	77	0.551	0.110	0.0001 – 0.05
Trichloroethane	77	1.06	0.161	0.0001 – 0.04
m-Xylene	77	0.708	0.113	0.0001 – 0.001
m-Xylene	77	0.527	0.0703	0.001 – 0.05

Source: EAB Control Cost Manual, 3rd ed. Research Triangle Park, NC: U.S. Environmental Protection Agency, February 1987. (EPA/450/5–87–001A; NTIS: PB 87–166583/AS), pp. 4–10.
[a]Each isotherm is of the form:
  m_e (lb adsorbate/lb carbon) = ap^b
  where p = partial pressure (psia)
Data are for adsorption on Calgon type "BPL" carbon (4 × 10 mesh).
[b]Equations should NOT be extrapolated outside these ranges.

Solution: First, we calculate the inlet partial pressures:

(1) (60 ppmv/1,000,000)(14.696 psia) = 0.000882 psia
(2) (2000 ppmv/1,000,000)(14.696 psia) = 0.0294

In Table 6.3, notice that two partial pressure ranges are given for m-xylene. One of these partial pressures falls in the first range; the other, in the second. Substituting these pressures and the corresponding isotherm parameters into Equation 6.17, we obtain:

(1) m_e = $0.708(0.000882)^{0.113}$ = 0.320 lb/lb carbon

(2) m_e = $0.527(0.0294)^{0.0703}$ = 0.411 lb/lb carbon

Interestingly, although the partial pressure for (1) is over 30 times the pressure for (2), its equilibrium adsorptivity is only 30% higher.

In many (if not most) cases, however, the inlet organic is a mixture of compounds, not a single constituent. How can we predict the equilibrium adsorptivity of such mixtures? Unfortunately, there is no definitive theory to fall back on. Some suggest figuring the mixture's adsorptivity based on the adsorptivity of the *least adsorbable* compound. Another expert states that carbon tends to adsorb larger and more polar molecules before the smaller, less polar constituents. Thus, a carbon bed would prefer the higher molecular weight compounds in a mixture over the lower. (That is, the larger molecules would fill up the carbon pores first, leaving the rest of the adsorption "sites" to the others.)[6] In addition, due to its high polarity, water is *very* easily adsorbed and, if present in large enough concentrations, will occupy all the active sites, leaving the organics to pass through the bed. This is why, to achieve optimum adsorption efficiency, the moisture content of the inlet waste gas stream must be kept to a minimum – normally, to < 50-60% relative humidity.

In adsorbers used for emission control, however, the amount adsorbed on the carbon rarely reaches the equilibrium level. Once the carbon bed reaches equilibrium with the adsorbate in the waste gas, its adsorption capacity is exhausted and the outlet concentration increases to the inlet level (i.e., it "breaks through"). But, to meet an emission regulation, the outlet concentration must be below a certain level during the entire time of the adsorption cycle. Hence, an excess amount of carbon is installed in the system to make sure that enough of the incoming adsorbate is captured to maintain the outlet concentration.

The two types of adsorbers commonly used to control emissions are *fixed-bed regenerable* and *modular* (or "canister") units.[7] Fixed-bed adsorbers can accommodate a wide range of flowrates and adsorbate concentrations. They are operated either *intermittently* or *continuously*. Intermittent units adsorb for a period of time (e.g., 8 hours) and then shut down for desorption. During desorption, the captured adsorbate is removed from the bed, most commonly via steaming. Following steaming, the adsorbate-steam vapor blend is condensed and then separated in a decanter into organic and aqueous layers. (However, if the organic is water-soluble, it must be separated via distillation.) Meanwhile, the desorbed bed is dried and cooled via a fan and readied for the next adsorption cycle.

Continuously operated fixed-bed systems operate similarly, except that an extra carbon bed (or beds) is included here to accommodate the inlet waste gas, while the other bed is being desorbed. Because the desorption cycle requires 1 to 1-1/2 hours, each bed must be large enough to handle the entire gas stream while the other bed is off-line. If the desorption time is considerably shorter than the adsorption time, it may be more economical to have two or more beds adsorbing, with only one desorbing. This arrangement can reduce the total system carbon requirement. (Reference 7 discusses this in more detail.) Figure 6.3 shows a schematic of a fixed-bed, steam-regenerated adsorber unit. Figure 6.4 is a photograph of a fixed-bed adsorber in use.

As Figure 6.3 shows, some waste gases require pretreatment before entering the adsorber to remove unwanted particulate and/or moisture. The former is captured by a filter (e.g., a baghouse), while a precooler is often used to reduce the inlet moisture content.

Unlike their regenerable brethren, modular adsorbers are discarded following completion of the adsorption cycle. Mostly they are used to control low-volume streams (1500 acfm or less), such as conservation vents on storage tanks. A typical modular adsorber consists of a corrosion-resistant steel or high-density polyethylene drum, activated carbon, inlet and outlet connections, a rain shield, condensate drain, and a stainless steel screen distributor.[8] A modular adsorber is shown in Figure 6.5.

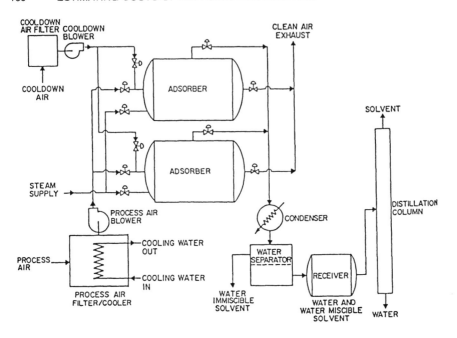

Figure 6.3 Schematic of fixed-bed carbon adsorber (courtesy Hoyt Corporation).

Because modular adsorbers contain little or no instrumentation, deciding when to shut down the unit can be a tricky proposition. Fortunately, most vendors provide an indicator to show when the carbon is near saturation, so that the unit can be replaced in time to prevent an exceedance of the emission regulation.

With regenerable adsorbers, however, a method is used to regulate the adsorption/desorption of the carbon beds to ensure that the outlet concentration is not exceeded. This method may be: (1) manual switchover, (2) a timer, or (3) an emission monitor. The first two are self-explanatory. Emission monitors are of two types: *electrolytic* or *photoionization*. Electrolytic monitors employ sensors that detect the gas concentration present via the amount of electric current that passes through a solid state material. In photoionization monitors, organic molecules passing through a sensor absorb ultraviolet light, which causes them to ionize. The migration of the charged molecules to an electrode creates a current, the magnitude of which is proportional to the measured organic concentration.[6]

Finally, what happens to the activated carbon when it can no longer be used? The carbon in regenerable adsorbers is replaced with fresh carbon. The spent carbon is reactivated by the carbon vendor and reused. But with modular adsorbers, the carbon—along with the drum, fittings, etc.—is usually disposed of. In pre-RCRA times, the adsorbers would simply be dumped at a municipal landfill. Not any more. As discussed in Chapter 4, the latest RCRA regulations severely restrict the kinds of substances that can be landfilled. Even if a pro-

Figure 6.4 Fixed-bed regenerable carbon adsorber fabricated of stainless steel (courtesy Amcec Corporation).

scribed ("listed") compound is present in small concentrations on an inert material (such as carbon), ordinary landfilling may be *verboten*. In such cases, the canisters would have to be incinerated or sent to RCRA-type landfills, where disposal costs (exclusive of transportation) are circa $250/ton.

Sizing Procedure

The primary sizing parameter for adsorbers is the *total carbon requirement* (W_c, lb). This parameter directly determines the cost of the initial carbon charge, and indirectly determines the size and number of the adsorber vessels and the auxiliaries (such as the system fan). The carbon requirement, in turn, incorporates several system variables: (1) adsorption time, (2) waste gas volumetric flowrate, (3) inlet and desired outlet adsorbate mass loadings, and (4) the working capacity (m_w) of the carbon.

We've already discussed the first three variables. The fourth, working capacity, accounts for the spare carbon capacity needed to keep the outlet concentration below the allowable level (e.g., 25 ppmv). Depending on the adsorbate, carbon, and operating conditions, the working capacity can range from less than 20% to nearly 100% of equilibrium capacity. However, the rule of thumb used by most adsorber vendors is 50% of the equilibrium capacity (m_e). For instance, if m_e is 0.20 lb adsorbate/lb carbon, the working capacity would be

Figure 6.5   Modular carbon adsorbers (courtesy Tigg Corporation).

0.10 lb/lb. In effect, this would *double* the carbon requirement predicted by the adsorption isotherm alone.

Alternatively, one may predict the working capacity using a sophisticated model that takes into account the above input variables plus other mass transfer parameters (e.g., the superficial bed velocity). Models like this one can yield more accurate predictions of the working capacity than the 50% rule of thumb. However, the calculations these models require are complex and require the use of at least a microcomputer.

How are the working capacity and the other three variables used to estimate the carbon requirement? First, for *intermittently* operated adsorbers:[7]

$$W_{ci} = (w_{ad})(t_a)/m_w \qquad (6.18)$$

where W_{ci} = carbon requirement for an intermittent adsorber (lb)

w_{ad} = adsorbate inlet loading (lb/hr)

t_a = adsorption time (hr)

m_w = working capacity (lb adsorbate/lb carbon)

For the special case where m_w is 50% of m_e (equilibrium capacity), substitute "$0.50m_e$" for m_w.

Recall that an intermittent adsorber is shut down at the end of the adsorption cycle for desorption. Although most intermittent adsorbers are small-capacity, single-bed units, they can contain several beds, should the total gas flowrate exceed approximately 10,000 acfm.

Because *continuous* units must have at least two beds (one on-line, one off), their carbon requirements are higher. In general, we can show that:[7]

$$W_c = W_{ci}(1 + N_d/N_a) \qquad (6.19)$$

where W_c = weight of carbon in *continuously* operated adsorber (lb)

W_{ci} = weight of carbon in intermittently operated adsorber *treating the same waste gas stream* (lb)

N_a, N_d = number of beds adsorbing and desorbing, respectively, at any given time

The following expression relates Na and Nd to the adsorption and desorption times:[7]

$$N_a = (t_a)(N_d)/(t_d) \qquad (6.20)$$

Example: In the previous example, we calculated the equilibrium adsorptivity for *m*-xylene at a 2,000 ppmv inlet concentration to be 0.411 lb/lb carbon. Assuming a 77°F operating temperature, 1-atmosphere operating pressure, a working capacity of one-half the equilibrium capacity, an inlet flowrate of 9000 scfm, and adsorption and desorption times of 4 and 1.5 hours, estimate the total carbon requirement for a regenerable adsorber unit.

Solution:

w_{ad} (inlet mass rate) = $(2000/1,000,000)(9000)(60)(0.274 \text{ lb/ft}^3)$
= 296 lb/hr

m_w (working capacity) = $(0.50)(0.411)$ = 0.206 lb/lb

Substituting these values and the adsorption time into Equation 6.18, we get:

$$W_{ci} = (296)(4)/(0.206) = 5740 \text{ lb of carbon}$$

But this is just the carbon in the adsorbing vessels. To determine the system carbon requirement, we need to know the number of adsorbing and desorbing vessels. The number of vessels in the system depends on the waste gas flowrate and bed velocity, as well as the adsorption and desorption times. A key parameter not mentioned above is the maximum vessel diameter. Due to transportation constraints, the largest vessel diameter that can be shop-fabricated is 12 ft.[7] Therefore, the largest flowrate (Q_{max}) that a 12-ft-diameter, vertically erected vessel can handle would be:

$$Q_{max} = (\pi/4)(80 \text{ ft/min})(12)(12) = 9050 \text{ acfm}$$

As this gas flowrate is 9000 scfm, one adsorbing vessel would (barely) suffice. Had the flow exceeded 9050, two vessels would have been needed. Finally, from Equation 6.20:

$$N_d = (1)(1.5)/(4) = 0.375 \text{ or } 1$$

Hence, we have a "one (vessel) on, one off" system. Equation 6.19 gives us the total carbon requirement for the system:

$$W_c = 5740(1 + 1/1) = 11,500 \text{ lb}$$

Equation 6.18 seems to tell us that, to minimize the total carbon requirement (and the adsorber equipment cost), we would need to keep the adsorption time as low as possible—i.e., no longer than the desorption cycle time—1.5 hours or so. But Equations 6.19 and 6.20 tend to contradict this hypothesis. For instance, Equation 6.20 tells us that, if the desorption and adsorption times were equal, then $N_a = N_d$. In such a case, the total carbon requirement would be *twice* the carbon needed in an intermittently operated system. However, if t_a were higher—say, twice t_d—then $N_a = 2N_d$ and $W_c = 1.5W_{ci}$. Although W_{ci} would be higher in the second case, the total carbon requirement would be a lower multiplier of W_{ci}. So, lengthening the adsorption cycle can work to reduce the total carbon requirement, as well as to increase it.

But there are considerations other than equipment cost. Shorter adsorption times require the bed to be desorbed more frequently. This increases the steam cost, for every time a bed is steamed the *entire* adsorber—carbon, vessel, and all—has to be heated, dried, and cooled. It makes more sense to desorb less frequently to keep down utility costs (steam, cooling water, and electricity). Finally, too-frequent desorption tends to shorten the carbon life. As a first estimate, therefore, assume an adsorption cycle time of 8 to 12 hours.[7]

Finally, much of the above discussion would not apply to the sizing of *modular* adsorbers, for these units are not regenerated. In this sense, they are similar to the intermittently operated regenerable adsorbers. Hence, the canister carbon requirement can be calculated from Equation 6.18, with m_w set at $\geq 0.50m_e$ and t_a selected at some arbitrary period (e.g., one week). Alternatively, one could calculate the time a modular adsorber could be run until the outlet concentration would begin to exceed the allowable emission limit. The following example illustrates the latter approach:

Example: A 100-scfm air stream containing 100 ppmv of toluene at 77°F is to be controlled with a modular adsorber. If the adsorber contains 200 lb of carbon and the working capacity is set at 50% of the equilibrium adsorptivity, estimate the time the unit can stay on-line before needing replacement. Repeat the calculation assuming that $m_w = 0.75m_e$.

Solution: At this inlet concentration, the partial pressure of toluene is:

$$(100/1,000,000)(14.696) = 0.00147 \text{ psia}$$

From Table 6.3, we obtain the toluene isotherm parameters: a = 0.551; b = 0.110. Substituting into Equation 6.17: $m_e = 0.551(0.00147)^{0.110} = 0.269 \text{ lb/lb carbon}$.

And: $m_w = (0.50)(0.269) = 0.135 \text{ lb/lb}$

The toluene inlet loading is:

$$W_{ad} = (100 \text{ scfm})(60)(100/1,000,000)(0.238 \text{ lb/ft}^3)$$
$$= 0.143 \text{ lb/hr}$$

Substituting these values into Equation 6.18 and rearranging:

$$t_a = (200 \text{ lb carb.})(0.135 \text{ lb/lb carb.})/0.143 \text{ lb/hr})$$
$$= 189 \text{ hr or about 8 days}$$

Finally, if we assumed a working capacity of 75% of m_e:

$$t_a = (189)(0.75/0.50) = 284 \text{ hr (almost 12 days)}$$

Finally, the pressure drop through an adsorber bed depends upon the bed thickness, superficial bed velocity, type of carbon, etc. For one commonly used carbon (Calgon's "PCB," 4 × 10 mesh), the pressure drop-velocity relationship is:[7]

$$\Delta P_b/l_b = 0.0368v_b + 0.000111(v_b)^2 \qquad (6.21)$$

where
l_b = bed thickness (ft)
ΔP_b = bed pressure drop (in. w.c.)
v_b = superficial bed velocity (ft/min)

The bed velocity typically ranges from 75 to 100 ft/min, depending on the system.

If pressure loss data for the carbon are unavailable, estimate the *total system* pressure drop for a fixed-bed regenerable adsorber at 8 to 10 in. w.c.[9] This pressure drop corresponds to an average bed velocity of 80 ft/min, and includes other pressure losses through the adsorber unit. (Pressure drops through the ductwork and other auxiliaries would be extra, however.)

While the bed pressure drop for the bed cooling/drying fan would be the same, the air flowrate and operating time would be different. Example: for a 30-min cooling/drying cycle, the air requirement would be approximately 3 to 3.5 scfm/lb of carbon in the bed.[7]

For modular adsorbers, the pressure loss varies with the adsorber capacity, expressed by the gas flowrate. Measured at their minimum (flowrate) capacities, the adsorber pressure drops range from 5 to 11.5 in. w.c., with 6 to 8 in. w.c. being typical.[8]

Costing Procedure

Equipment Costs

This section contains prices for both fixed-bed regenerable and modular (canister) adsorbers.

Fixed-bed regenerable. Like most other types of control devices, fixed-bed adsorbers can be sold either as "packaged" or "custom" units, depending on the size of the adsorber, as measured by the total carbon weight, W_c. First, we obtained vendor quotes for carbon steel, steam-regenerated *packaged* adsorbers.[9] These prices (in **December 1988** dollars) fit the following regression curve:

$$P(\$) = 129(W_c)^{0.848} \qquad (6.22)$$

where $350 \leq W_c, \text{lb} \leq 14,000$

(These carbon weights roughly correspond to waste gas flowrates of 700 and 25,000 scfm, respectively.) Each of these prices includes the adsorber vessels, carbon, condenser, decanter, system fan and motor, bed cool-down fan and motor, instruments and controls, and internal ducting. The cost of neither a boiler (to provide regeneration steam) nor an emission monitor is included. A monitor (specifically, a photoionization detector) would add $10,000 to $12,000 to the price of a system. The monitor's operating and maintenance costs can be considered negligible, however.

The same kinds of equipment are included in the prices of the *custom* fixed-bed regenerable adsorbers. Because these are custom adsorbers, each of their components is individually designed and built to accommodate the needs of the emission source. Moreover, the units may or may not include the same equipment as the packaged adsorbers.

Based on the custom prices in reference 7 (which were escalated from April 1986 to **June 1988** dollars), the following correlation was developed:

$$P(\$) = 32.8(W_c)^{0.860} \qquad (6.23)$$

where $14,000 \leq W_c \leq 222,000$

(The flowrates corresponding to these carbon weights range from >25,000 to 500,000 scfm.) Each of the above prices includes the same equipment as do the packaged adsorber prices.

Lastly, for those interested in making quick ("order-of-magnitude") estimates, an anonymous adsorber vendor has kindly provided the following installed costs (**summer 1988** dollars) for carbon steel adsorbers:

Waste Gas Flowrate (scfm)	Total Capital Investment ($/scfm)
1 to 2000	50 to 75
30,000 to 500,000	20 to 25

These costs include all the costs for purchasing and installing the adsorber equipment, except for the customer-supplied auxiliaries, such as ductwork. If the above adsorbers were fabricated of 316 stainless steel or Hastelloy, the prices would be twice the $/scfm values shown.

Modular adsorbers. Based on vendor prices, the following price correlation was developed for modular adsorbers:[8]

$$P(\$) = 4.66(W_c)^{0.968} \tag{6.24}$$

where $110 \leq W_c, \text{lb} \leq 5,700$

Escalated from August 1987 to **June 1988** dollars, each of these prices includes a vessel (a carbon steel or polyethylene drum), activated carbon, flow distributor (stainless steel screen), inlet/outlet connections, rain shield, and condensate drain. Options include a visual saturation indicator ($40) and an aluminum flame arrestor ($400). Customers buying 10 or more units of a given size would be given up to a 5% discount from the above prices.

Total Capital Investment

Packaged regenerable and modular adsorbers. Because these units come equipped to be virtually "plugged into" emission sources, the costs for installing them are minimal. (See similar discussion for thermal and catalytic incinerators earlier in this chapter.) A reasonable installation cost estimate would be 25% of the PE (purchased equipment cost). Thus:

$$TCI = 1.25(PE) \tag{6.25}$$

Custom regenerable adsorbers. Because the components of custom adsorbers are part vendor-fabricated and part field-assembled, their installation factors are higher. The average composite installation factor listed in Table 2.2 for carbon adsorbers would apply here, i.e.:

$$TCI = 1.61(PE) \tag{6.26}$$

Annual Costs

The total annual cost (TAC) consists of *direct* and *indirect* annual costs, offset by *adsorbate recovery credits*, where applicable. *Direct* annual costs for adsorbers include labor (operating, supervisory, and maintenance), maintenance materials, electricity, steam, cooling water, replacement carbon, and solid waste disposal. Of these, only electricity and solid waste disposal apply to modular adsorbers. The unit prices for these costs will vary by installation. Nonetheless, general consumption figures can be given. First, Chapter 2 (Table 2.5) recommends these values:

- Operating labor: 0.5 hr/shift
- Supervisory labor: 15% of operating labor
- Maintenance labor: 0.5 hr/shift
- Maintenance materials: 100% of maintenance labor

For the other direct annual costs, the following are suggested:

- Electricity: Compute the system fan electricity requirement (in kW) from Equation 2.2, inputting the waste gas flowrate (in *actual* ft^3/min) and the system pressure drop discussed above for fixed-bed and modular adsorbers, respectively. (Recall that the system pressure drop is the sum of the pressure losses through the adsorber unit and the auxiliary equipment (e.g., ductwork).) The power requirement for the bed drying/cooling fan would be figured similarly, inputting the *bed* pressure drop in Equation 2.2. However, the gas flowrate and operating hours would be different. (See reference 7 for a fuller discussion of this.)

- Steam: The steam requirement is 3 to 4 lb/lb adsorbate captured.[9] This is an average value, however, measured over the length of the entire bed steaming cycle (typically, 1 hour in duration). The peak steam demand occurs during the first 10 to 15 minutes of the cycle, and is 2 to 2.5 times the average demand.

- Cooling water: The cooling water is consumed by the condenser to condense the water vapor-adsorbate mixture. The water usage is a function of the steam usage and the water temperature rise—typically, 30 to 40°F.[7] Assuming a 35°F water temperature rise and 1000 Btu/lb latent heat for the steam-adsorbate mixture, we can derive the following:

$$G = 3.43S \tag{6.27}$$

where   G = cooling water usage (gal/yr)
             S = steam usage (lb/yr)

- Carbon replacement: The carbon has a shorter life than the rest of the adsorber unit—typically, 3 to 5 years compared to 10 years for the equipment.[9] Hence, the carbon replacement cost (C_{cb}) must be calculated separately. The following equation, analogous to Equation 6.15 above, was developed for that purpose:

$$C_{cb}(\$) = (P_{cb} + P_{cbl})(CRF_{cb}) \qquad (6.28)$$

where
P_{cb} = initial price of system carbon (\$), including sales taxes and freight
P_{cbl} = labor cost to replace carbon (\$)
CRF_{cb} = capital recovery factor for useful life of carbon and interest rate in question

The carbon replacement labor would vary according to the amount replaced, labor rates, etc. Reference 7 gives a figure of \$0.05/lb to replace a 50,000-lb carbon charge. Compared to a typical carbon price (\$2.00/lb), this labor cost would be negligible, however.

• Solid waste disposal: As discussed earlier, spent canister adsorbers must be disposed of—drum, carbon, fittings, et cetera. Depending on the toxicity of adsorbate(s) captured, the canisters may be shipped either to a municipal or a RCRA landfill. Disposal costs at the former range from \$10 to \$40/ton; at the latter, upwards of \$250/ton. Depending on their capacity, modular adsorbers weigh from 1/4 ton to over 4 tons apiece.

Indirect annual costs include capital recovery, overhead, property taxes, insurance, and administrative charges. As Chapter 2 recommends, figure the last three items at 4% of the TCI and calculate the overhead at 60% of the sum of all labor and maintenance materials costs.

To compute the capital recovery cost for fixed-bed regenerable adsorber systems (CRC_s), use this equation:

$$CRC_s = (TCI - P_{cb} - P_{cbl})(CRF_s) \qquad (6.29)$$

The CRF_s is based on the annual interest rate and the system lifetime (typically, 10 years). The P_{cb} and P_{cbl} terms must be deducted from the TCI to avoid double-counting in the capital recovery calculation.

The capital recovery cost for modular adsorbers would be figured differently, as their lifetimes are usually less than one year. However, the ducting and other auxiliaries would last several years. Thus, we can treat the cost of the canister as an operating expense, while amortizing the rest of the system cost (e.g., ductwork) over its lifetime—typically, 10 years.

The *recovery credits* apply to those adsorbates captured by the system, condensed, and separated that have value either as recyclable or resalable commodities. Not all adsorbates can be reused, regardless of their purity.

Others may be reusable, but the costs to separate them from the steam conden-sate and other adsorbates may exceed their value. Finally, some adsorbates may only have value as fuel. But if an adsorbate has value, one can estimate the amount recovered (R, lb/yr) as follows:

$$R = (w_{ad})(E)(H) \qquad (6.30)$$

where E = fraction of inlet adsorbate removed during adsorp-tion cycle

w_{ad} = inlet adsorbate rate (lb/hr)

H = annual operating hours (hr)

The control efficiency here is figured based on the difference between the inlet adsorbate rate and the outlet rate *figured at the breakthrough concentra-tion*. This gives a conservative estimate of the fraction of adsorbate removed during the adsorption cycle.

Example: Estimate the capital and annual costs of the adsorber we sized in the previous example. Assume that the ductwork and other auxiliaries are in place, and that the unit operates 8000 hr/yr, achieving 95% removal of the *m*-xylene.

Solution:

1. *Equipment cost:* The total carbon requirement we calculated in the previous example was 11,500 lb. Substituting this value in Equation 6.22, we get:

$$P(\$) = 129(11,500)^{0.848} = \$358,000$$

2. *Total capital investment:* As this is a packaged system, the installa-tion cost is low – 1.25 times the purchased equipment cost (PE). As the ductwork, etc., are already in place, the PE consists of the adsorber equipment cost plus the cost of taxes and freight. (Instrumentation is already included in the equipment cost). Thus:

$$TCI = 1.25(PE) = (1.25)(1.08)(\$358,000) = \$387,000$$

3. *Direct annual costs:* Based on the 8000 hr/yr = 1000 shifts/yr operating factor and the unit prices shown below, we have:

- Operating labor = (0.5 hr/sh.)(1000)($10/hr)
  = $5000
- Supervisory labor = (0.15)($5,000) = $800
- Maintenance labor = (0.5 hr/sh.)(1000)($11/hr)
  = $5,500

(*Note:* the maintenance labor rate is figured at 110% of the operating labor rate, as Chapter 2 recommends.)

- Maintenance materials: $5,500
- Electricity: Estimate the system pressure loss at the high end — 10 in. w.c. — to cover the cooling fan power consumption. From the previous example, gas flowrate is 9000 scfm at 77°F or 9120 acfm. Assuming a power cost of $0.05/kW-hr and a 65% fan-motor efficiency, we have:

Power cost = (0.746)(0.0001575)(1/0.65)(10)(9120) × (8000)($0.05)
            = $6600

- Steam: From the last example, the *m*-xylene inlet rate was 296 lb/hr. Figuring a (average) steam usage of 3.5 lb/lb adsorbate and a $6.00/1000 lb steam price, we have:

Steam cost = (3.5)(296)(8,000)($0.006) = $49,700

- Cooling water: Based on Equation 6.27 and a $0.20/1000 gal cooling water price:

Cooling water cost = (3.43)(3.5)(296)(8,000)($0.0002)
                   = $5,700

- Carbon replacement: Based on a 3-year carbon life, a 10% annual interest rate, and a negligible carbon replacement labor cost, we have:

Carbon replacement cost = (11,500 lb)($2.00/lb)(1.08)(0.4021) = $10,000

(*Note:* "0.4021" is the capital recovery factor for a 3-year life and a 10% annual interest rate.)

Summing the above costs:
Subtotal, directs: $88,800

4. *Indirect annual costs:*

- Property taxes, insurance, administrative charges:

PT, I, & AC = (0.04)($387,000) = $15,500

- Overhead = (0.60)($5,000 + $800 + $5,500 + $5,500) = $10,100
- Capital recovery: Based on a 10-year system life, a 10% annual interest rate, and Equation 6.29:

CRC = ($387,000 - $24,800)(0.1628) = $59,000

(*Note:* "0.1628" is the capital recovery factor for a 10-year life and a 10% annual interest rate.)

Summing these three costs:
Subtotal, indirects: $84,600

5. *Recovery credit:* Assuming a recovered *m*-xylene resale value of $1.50/gal,[6] a density of 7 lb/gal, and a 95% collection efficiency, we have:

Recovery credit = (0.95)(296)(1 gal/7 lb)(8,000)($1.50/gal) = $482,100

6. *Total annual cost:*
$$\text{TAC} = \text{Directs} + \text{Indirects} - \text{Credit}$$
$$= \$88,800 + \$84,600 - \$482,100$$
$$= -\$309,000/\text{yr (rounded)}$$

Hence, this adsorber makes money for the source.

Clearly, the *m*-xylene recovery credit dominates the TAC. This credit is, in turn, extremely sensitive to the value selected for the recovered *m*-xylene. Had we selected a value of, say, $0.50/gal, the credit would have been only $161,000 and the TAC would have been a net cost of approximately $12,000. By comparison, the various direct and indirect costs are unimportant. All of this highlights the need to select the adsorbate price and other parameters very carefully.

REFRIGERATED CONDENSERS

Description

Refrigerated condensers are used to remove volatile organic compounds from waste gas streams by cooling the streams to temperatures ranging from 40 to -110°F. The low recovery temperatures are achieved by coupling the condensers to single or multi-stage ("cascade") refrigeration units.

Most of these refrigeration units operate via the basic mechanical compression cycle, which consists of the following steps: (1) heat absorbed from the waste gas in the condenser evaporates the refrigerant; (2) the refrigerant is then compressed to a higher temperature and pressure; (3) the refrigerant rejects sensible and latent heats in a condenser (at temperatures ranging from 0 to -40°F); and (4) after passing through an expansion valve, the refrigerant is vaporized, completing the cycle.

Refrigerated condensers are categorized based on the type of condenser used to remove the VOCs—i.e., *surface* or *direct-contact*. In surface condensers, the waste gas and refrigerant do not contact each other. A surface condensing system usually consists of a dehumidifier (for moisture removal), a recuperative heat exchanger (to precool the waste gas), and a vapor condenser (in which the VOCs are removed). Refrigeration units are connected to the dehumidifier and the vapor condenser to provide the coolant needed to achieve the low temperatures in these devices.[10] Figure 6.6 is a schematic of a refrigerated

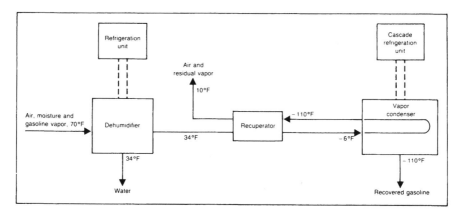

Figure 6.6 Schematic of surface-condensing refrigeration system. (Reprinted with permission from *Chemical Engineering*, May 16, 1983, pp. 95–98.).

surface-condensing system, while Figure 6.7 shows a typical refrigeration system used for vapor recovery.

While surface-condensing systems can achieve temperatures as low as −110°F, in direct-contact systems the lowest achievable temperature is 32°F. That is because temperatures lower than that would cause any moisture in the stream to freeze. In contact systems, the waste gas enters the bottom of a packed column and flows upward through a descending stream of organic absorbent — say, benzene. The absorbent captures the vapors, and the stream of air, moisture, and residual vapor exits the top of the column. Because the recovery temperature must be kept above 32°F, the efficiency of contact condensers is less than that of surface-condensing systems.[10] (See Figure 6.8.)

Sizing Procedure

A refrigeration unit is typically rated in "tons" — a measure of the amount of heat it can remove per unit time. (1 "ton" = 12,000 Btu/hr.) The heat removed is calculated from the following equation:[10]

$$q_r = q_a + q_v + q_c \qquad (6.31)$$

where q_a, q_v = sensible heat cooling of air and VOCs from inlet to outlet temperatures, respectively (Btu/hr)
q_c = heat evolved by condensing VOCs (Btu/hr)

The amount of VOCs condensed (w_c, lb/hr) depends on several variables:[10]

$$w_c = 0.00763(M_v Q'/T_1 + 460)[(y_1 - y_2)/(1 + y_1)] \qquad (6.32)$$

Figure 6.7   Surface-condensing vapor recovery unit (courtesy Edwards Engineering Corporation).

where M_v = molecular weight of VOCs (lb/lb-mole)
Q' = inlet volume of waste gas (actual gal/day)
T_1 = inlet temperature of waste gas (°F)
y_1, y_2 = inlet and outlet saturated concentrations of VOCs (moles VOCs/mole air)

The inlet and outlet VOC concentrations depend primarily on their vapor pressures at temperatures T_1 and T_2. Furthermore, Equation 6.32 is based on negligible water vapor content and constant-pressure operation at 1 atm.

To calculate the control efficiency of a refrigeration unit, simply divide w_c by the quantity of VOCs in the inlet waste gas, or:

$$\text{Efficiency} = w_c/w_{v1}$$
$$= (y_1 - y_2)/y_1 \qquad (6.33)$$

Most refrigeration units are designed to achieve a given outlet VOC concentration, at a constant waste gas inlet flowrate.

All of these variables determine the size (and price) of refrigeration units, explicitly or implicitly. Also determining the cost are such factors as mode of

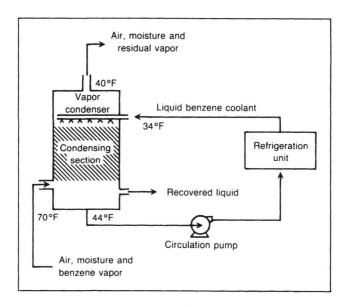

Figure 6.8 Schematic of contact-condensing refrigeration system. (Reprinted with permission from *Chemical Engineering,* May 16, 1983, pp. 95–98.).

operation (continuous or intermittent) and accessories included, such as a dehumidifier. However, for a given unit design and set of operating conditions—temperature, pressure, VOC molecular weight, etc.—the price of refrigeration units is usually correlated with Q′ (gal/day or scf/day), the inlet waste gas flowrate.

Costing Procedure

Equipment Costs

We obtained vendor prices (in **February 1988** dollars) for four surface-condensing refrigeration system designs.[11] All were designed for recovering hydrocarbon vapors at gasoline bulk terminals. Hence, the price correlations below should not be indiscriminately applied to other vapor recovery situations. Three of the units were designed for intermittent operation. (That is, they would be shut down for one hour each day for defrosting.) The fourth unit was designed for continuous operation. The prices for these units fit the following regression correlation:

$$P(\$) = aQ'^b \tag{6.34}$$

where Q' = refrigeration unit capacity (gal/day)
          a, b = regression parameters

The values for these parameters were:

Unit	Parameter a	b	Correlation Range (1000 gal/day)
A	1400	0.380	300 to 1200
B	4930	0.292	375 to 2000
C	296	0.517	400 to 2400
D	6.61	0.791	3200 to 4800

Prices provided by these correlations cover the entire packaged refrigeration unit, except for the heat transfer fluid. A typical unit requires 400 gallons. At a fluid cost of $3.50/gal, this would add $1400 to each of the above unit prices.

Units A, B, and C are operated intermittently, while unit D runs continuously. Each of the "A" units consists of a cascade refrigeration system using semi-hermitic refrigeration compressors and a direct expansion hydrocarbon condenser coil. The "B" units include the same equipment, plus a chiller pre-cooler, which, during the warm summer months, reduces frost formation on the low temperature hydrocarbon recovery coil. Both the "C" and "D" units contain the same equipment as the "B" systems, except that they use Carrier open-type compressors instead of the semi-hermitic compressors. A "D" unit is a "C" unit with an extra compressor set and condenser coil. The spare compressor and coil allow the unit to keep operating while the other coil defrosts. However, during this defrost cycle (one hour/day) the unit operates at reduced capacity. Finally, all four units are equipped with pressurized compressor enclosures and full instrumentation.[11]

When operated at its design capacity, each of these units can reduce the hydrocarbon concentration to 0.29 lb/1000 gal of waste gas inlet volume. At a hydrocarbon density of 7 lb/gal, this is equivalent to an outlet concentration of about 40 ppmv. If the inlet flowrate to a unit were *doubled*, however, the outlet concentration would increase to 0.67 lb/1000 gal (about 100 ppmv). Although the fraction of hydrocarbon removed will vary depending on the waste gas flowrate, composition, and other characteristics, overall removal efficiencies exceeding 90% are the norm.[11]

Total Capital Investment

Because nearly all refrigeration units are packaged, installation costs tend to be low compared to the installation costs for custom equipment. A vendor suggests an estimate of 10 to 15% of PE.[11] Using the high end of this range, we can say that:

$$TCI = 1.15(PE) \qquad (6.35)$$

Annual Costs

As with carbon adsorbers, the TAC of a refrigeration unit consists of *direct* and *indirect* annual costs, offset by *VOC recovery credits*. *Direct* annual costs for refrigeration units include labor (operating, supervisory, and maintenance), maintenance materials, and electricity. Although the unit prices for these items will vary according to the installation, general consumption figures can be given. First, Chapter 2 recommends these values:

- Operating labor: 0.5 hr/shift
- Supervisory labor: 15% of operating labor
- Maintenance materials: 100% of maintenance labor

A refrigeration system vendor[11] suggests the following:

- Maintenance labor: 0.5 hr/day
- Electricity: 1 to 1.5 kWhr/1000 gal of inlet waste gas

Indirect annual costs include capital recovery, overhead, property taxes, insurance, and administrative charges. As Chapter 2 recommends, figure the last three items at 4% of the TCI and estimate the overhead at 60% of the sum of all labor and maintenance materials costs.

To estimate the capital recovery cost, multiply the capital recovery factor by the TCI. When calculating the CRF, use the appropriate annual interest rate and a system life of 15 to 20 years.[11]

The *recovery credit* pertains to the VOC condensate (w_c, lb/hr) recovered by the refrigeration unit. The amount of this credit would be the product of the condensate's value and the quantity recovered, the latter being provided by this equation:

$$R = (w_c)(H) \tag{6.36}$$

where
    R = amount of condensate recovered annually (lb/yr)
    H = annual operating hours (hr)

There may be occasions where the value of recovered condensate would exceed the direct and annual costs of the refrigeration system. The next example illustrates such a case:

Example: A refrigeration system must be installed to control gasoline vapor emissions generated by the loading of tank trucks at a bulk terminal. The total volume of vapor displaced (equivalent to the waste gas inlet flowrate) is 4 million gallons/day. The refrigeration unit must operate continuously for 8400 hours/year. (During the rest of the year, the terminal is not in operation.) During this time, it must remove at least 90% of the gasoline vapors in the waste gas, to achieve an outlet concentration of 0.30 lb/1000 gal or less. Estimate the capital and annual costs of this unit, and the gasoline recovery credit, figuring the latter at $1.00/gal. (Assume 7 lb/gal for the density of gasoline.)

Solution:

1. *Total capital investment:* First, estimate the equipment cost, via Equation 6.34. Because this refrigeration system must operate continuously, a unit "D" must be selected. The correlation parameters for this unit are: a = 6.61; b = 0.791. Thus:

$$\text{Price of unit} = 6.61(4,000,000)^{0.791} = \$1,103,000$$

To this price we must add the costs of heat transfer fluid ($1400), freight (5%), sales taxes (3%), and auxiliary equipment. (Instrumentation is already included in the purchase price.) The auxiliary equipment consists of ductwork, costing an estimated $12,000. The purchased equipment cost is, therefore:

$$\text{PE} = (\$1,103,000 + \$1400 + \$12,000)(1.08) = \$1,206,000$$

Using the above-recommended installation factor, we have:

$$\text{TCI} = 1.15(\text{PE}) = \$1,390,000 \text{ (rounded)}$$

2. *Annual costs:* Based on this operating schedule (8,400 hr/yr = 1050 shift/yr = 350 days/year), the above consumption figures, and the prices below, we can estimate the following:

- Operating labor = (0.5 hr/sh.)(1050)($10/hr) = $5,250
- Supervisory labor = (0.15)($5,250) = $790
- Maintenance labor = (0.5 hr/day)(350)($11/hr) = $1900
  (*Note:* the maintenance labor rate has been figured at 110% of the operating labor rate, as Chapter 2 suggests.)
- Maintenance materials = (1.00)($1900) = $1900
- Electricity: To be conservative, figure this at the high end of the above range, or 1.5 kWhr/1000 gal of waste gas. At $0.05/kWhr, we have:

$$\text{Electricity cost} = (0.05)(1.5)(4,000,000/1000)(350) = \$105,000$$

Subtotal, direct costs: $114,840
The indirect annual costs are:

- Overhead = (0.60)($5,250 + $790 + $1900 + $1900) = $5900
- Property taxes, insurance, administrative charges:

$$\text{PT, I, \& AC} = (0.04)(\$1,390,000) = \$55,600$$

- Capital recovery: To be conservative, assume a 15-year life for the system. The capital recovery factor for this life and a 10% annual interest rate is 0.1315. Thus:

$$\text{CRC} = (0.1315)((\$1,390,000) = \$182,800$$

Subtotal, indirect costs: $244,300
To calculate the gasoline recovery credit, we need to know how much

the unit collects in a year. Based on an outlet loading of 0.30 lb/1000 gal of waste gas and a 90% removal efficiency, we have:

$$\text{Amount recovered} = (0.30)(.90/.10)(4,000,000/1000)(350)$$
$$= 3,780,000 \text{ lb/yr}$$

Assuming a condensate density of 7 lb/gal and a value of $1.00/gal:

$$\text{Recovery credit} = (3,780,000)(1/7)(\$1.00) = \$540,000$$

Finally, we add the direct and indirect annual costs and subtract from this sum the recovery credit, to obtain the total annual cost/credit (TAC):

$$\text{TAC} = \$114,840 + \$244,300 - \$540,000 = -\$181,000 \text{ (rounded)}$$

Based on these numbers, the bulk terminal owners would be well advised to install this refrigeration unit!

FLARES

Description

Originally, flares were designed and installed in refineries and chemical plants to provide intermittent disposal of waste gases during startups or process upsets. They also have been used to continuously combust unwanted by-products from crude oil production and other petrochemical processes. Only in recent years have flares been applied to the continuous control of air pollutants. Unsurprisingly, most of these applications have been in petrochemical industries, although lately flares have been used to burn methane, hydrogen sulfide, and other noxious gases emitted from such sources as landfills and sewage treatment plants.

Most flares fall into two broad categories: *elevated* or *enclosed ground*. Elevated flares are mounted on stacks at some horizontal and vertical distance from the process area, for reasons of safety (radiation exposure) and noise avoidance. Along with the flare itself, these units consist of an ignition system and instrumentation (pilot flame detector and flow sensor), a stack liquid seal (to prevent air infiltration and flame flashback), a knockout drum or mist trap (to remove entrained liquid).[12]

Depending on the type of material burned, one of three elevated flare designs is used: *smoking*, *smokeless*, or *endothermic*. Clean-burning gases (e.g., ammonia, hydrogen sulfide, and hydrocarbons with hydrogen/carbon ratios exceeding 25%) are burned in smoking flares. Smokeless flares — so-called because steam is injected into them to produce a clear exhaust — are used to combust organics with lower H/C ratios. Finally, to burn gases with low

Figure 6.9 Ground flare controlling landfill gases. (Courtesy Sur-Lite Corporation.)

heat contents (< 100 Btu/ft^3), endothermic flares are employed. Auxiliary heat must be added to the latter.[12]

Ground flares are quite different from elevated flares in several respects. For one thing, they can burn both gaseous and liquid wastes, which are combusted via several burners either in a refractory chamber or an open pit. For another, their designs are much more complex than those of elevated flares, even though they require neither a stack nor supports.[12] (A typical ground flare is shown in Figure 6.9.) Several basic factors determine the design of a flare: (1) the waste gas flowrate and composition; (2) flaring frequency; (3) solar radiation and weather conditions (especially wind velocity and direction); (4) flare stack siting limitations (due to locations of personnel, adjoining property, hazardous material storage, etc.); and (5) expansion capacity. Although the design must consider these five factors, each flare design is unique and must be "tailor-made" for the source.[13]

Sizing Procedure

The sizing (and pricing) of a flare is based on several factors, so that no single sizing parameter predominates. (Indeed, flare design is as much — or more — an art as it is a science.) Oenbring[13] presents a comprehensive yet succinct sizing procedure for elevated flares, which, with some adjustments, could be adapted to the sizing of ground flares as well. Oenbring's procedure can be boiled down to two steps:

1. *Calculating the flare tip diameter* (D_t, in.):

$$D_t = 13.54(Q/v_e)^{0.5} \qquad (6.37)$$

where Q = waste gas outlet flowrate (acfm)
                 v_e = flare gas exit velocity (ft/min)

The gas exit velocity is, in turn, a function of the flared gas sonic velocity and the appropriate Mach number (0.2 and 0.5 for continuous and intermittent flaring, respectively).

2. *Calculating the required distance from the flare flame center to the nearest point of allowable radiation intensity* (Z, ft):

$$Z = 0.282(fq/k)^{0.5} \qquad (6.38)$$

where q = heat content of waste gas (Btu/hr)
               k = allowable heat flux from flare flame (Btu/hr-ft^2)
                 = 600 (long-term exposure)
              f = fraction of flame heat radiated
                 = 0.25 (light gas); 0.4–0.5 (heavier gases)

Once Z is computed, we can calculate the flare stack height for any given distance from the stack base to the "point of allowable intensity." Oenbring provides a detailed example that illustrates this procedure (which is somewhat more complex than shown here).

Costing Procedure

Equipment Costs

Notwithstanding the sizing procedure just summarized, for convenience flare prices are commonly correlated against just the waste gas flowrate, in either lb/hr or scfm. This section will present prices for elevated flares (self-supporting, guyed, and derrick), ground flares, and "candle flares" (a ground flare design variation).

Elevated flares. We obtained prices for elevated flares from reference 12, and escalated them to **June 1988** dollars. These costs fit the following correlation:

$$P(\$) = a(Q'')^b \qquad (6.39)$$

where Q'' = waste gas flowrate (lb/hr)
         such that $2{,}500 \le Q'' \le 250{,}000$
    a, b = correlation parameters

These prices include the flares (self-supporting), ladders, platforms, knock-out drums with seals, and stacks high enough (approximately 40 ft) to ensure grade-level radiation no greater than 1500 Btu/hr-ft^2. Values for these parameters are:

	Correlation Parameters	
Waste Gas Heat Content	a	b
Low	1430	0.256
High	471	0.398

The "low" heat content flares would be *endothermic* types, while the "high" could be either *smoking* or *smokeless*, depending on the waste gas composition.

Again, these prices are for self-supporting flares. To estimate the prices of *guyed* flares (100 ft tall) and *derrick* elevated flares (200 ft), reference 12 suggests multiplying the Equation 6.39 prices by the following factors:

Flare Type	Multiplier
Guyed	1.80 (@ Q'' = 2500 lb/hr)
	to
	1.30 (@ Q'' 250,000 lb/hr)
Derrick	$47.9(Q'')^{-0.214}$

Ground flares. A flare vendor provided costs for ground flares.[14] These flares have been designed primarily for burning gases generated by landfills and sewage treatment plants. The units are packaged and include full instrumentation. To sustain combustion, the units require waste gas heat contents of 550 to 600 Btu/scf. Their prices (**December 1988** dollars) fit the following correlation:

$$P(\$) = 23{,}600 + 21.4Q \qquad (6.40)$$

where Q = waste gas flowrate (scfm) such that $150 \le Q \le 4000$

"Candle" flares. Finally, we obtained costs for "candle" flares from another vendor (who elected anonymity). Each of these flares is essentially a waste gas

burner (metal) with an ignition system (auto pilot type). The unit is self-contained and instrumented and is installed at ground level. (The flare itself is set 10–12 ft above grade.) Candle flare prices (**December 1988** dollars) fit this equation:

$$P(\$) = 13,100 + 767D_t \qquad (6.41)$$

where D_t = flare tip diameter (in.) such that $2 \le D_t \le 10$

Total Capital Investment

Elevated flares. As nearly all elevated flares are custom-built, the installation factor given in Chapter 2 applies. That is:

$$TCI = 1.92(PE) \qquad (6.42)$$

where   PE = purchased equipment cost, as defined in Chapter 2

Ground flares. Conversely, most ground flares are packaged units, so that their installation costs would be lower. One flare vendor estimates the installation cost for ground, "candle," and related flares at 20 to 25% of the PE.[14]

Annual Costs

As with the other control devices covered in this and the previous chapter, the total annual cost (TAC) for flares is the sum of *direct* and *indirect* annual costs. The *direct* costs consist of maintenance labor and materials, fuel (natural gas), steam (where applicable), and electricity. (As flares operate automatically, either on a continuous or an as-needed basis, they require neither operating nor supervisory labor attention.) Although the prices of these commodities will vary from site to site, the following general consumption figures can be given:

- Maintenance labor: 0.5 hr/shift, as Chapter 2 suggests
- Maintenance materials: 100% of maintenance labor, again from Chapter 2
- Fuel: Both elevated and ground flares require natural gas. The former need it to supplement ("assist") the heat content of low-Btu waste gases. Although no assist gas is needed with ground flares, gas is needed for burner pilots and purging. The following equations provide estimates of these gas requirements:[12]

$$\text{Elevated: } F_{ef} = 2.18(Q'')^{0.934} \qquad (6.43)$$

$$\text{Ground: } F_{gf} = 0.0701(Q'')^{0.257} \qquad (6.44)$$

where

$$F_{ef}, F_{gf} = \text{fuel gas requirements for elevated and ground flares (million Btu/hr)}$$

$$2500 \leq Q'' \text{ lb/hr} \leq 250,000$$

- Steam: Where steam is required in smokeless or other flares requiring steam injection, the steam usage can be estimated at 0.6 lb/lb of waste gas.[12]
- Electricity: Power is consumed by the flare system fan, primarily. For elevated flares, the system pressure drop depends upon the size of the flare, knockout drum, and length and diameter of the piping, as well as the stack height. This pressure drop can be as high as 60 in. w.c.[12] For ground and candle flares, the pressure drops are lower, though again they would vary by installation. As estimating rules of thumb, use 1 to 2 in. w.c. for ground flares, and 0.5 to 1 in. w.c. for candle flares.

Indirect annual costs for flares include overhead, capital recovery, and the usual property taxes, insurance, and administrative charges. As usual, the last three can be estimated at 4% of the TCI. Figure overhead at 60% of the maintenance labor and materials costs. Finally, calculate the capital recovery by multiplying the TCI by the capital recovery factor. As Chapter 2 suggests, compute the CRF based on a 15-year life.

Example: A municipal landfill generates 2500 scfm of waste gas (mostly methane). Estimate the capital and annual costs of a ground flare sized to combust this waste gas, if the flare operates 8000 hr/yr.

Solution:

1. *Total capital investment:* First, calculate the flare equipment cost using Equation 6.40:

$$\text{Equipment cost} = 23,600 + 21.4(2500) = \$77,100$$

The purchased equipment (PE) cost would be this cost plus sales taxes and freight, as instrumentation is already included in the price. Thus:

$$PE = (1.08)(\$77,100) = \$83,300$$

As this would be a packaged flare, estimate the TCI as:

$$TCI = (1.25)(PE) = \$104,000 \text{ (rounded)}$$

2. *Total annual cost:* Calculate the direct annual costs using the factors suggested above and the prices listed below. Thus:

- Maintenance labor = ($11/hr)(0.5 hr/sh.)(1000 sh/yr) = $5500
- Maintenance materials = maintenance labor = $5500
- Fuel: As the waste gas heat content is high enough to be burned without any assist gas, the only gas required will be that for the pilots and purging. Using the gas usage given by Equation 6.44 and a $5.00/million Btu price, we calculate the following:

$$\text{Fuel cost} = (\$5.00)(0.0701[2{,}500]^{0.257})(8{,}000) = \$20{,}900$$

- Electricity: Assuming a 2-in. pressure drop, a 65% combined fan-motor efficiency, and a $0.05/kWhr electricity price, we would get the following:

$$\text{Electricity cost} = (\$0.05)(0.746)(0.0001575/0.65)(2)(2500)(8000)$$
$$= \$400$$

Summing the direct costs:
Subtotal, directs: $32,300

The indirect costs are:

- Overhead = (0.60)($5500 + $5500) = $6600
- Property taxes, insurance, and administrative charges:

$$\text{PT, I, \& AC} = (0.04)(\$104{,}000) = \$4{,}200$$

- Capital recovery: This is the product of the TCI and the CRF. Estimate the CRF using a 15-year system life, as suggested above, and a 10% annual interest rate. Thus:

$$\text{CRF} = 0.1315 \text{ and CRC} = (0.1315)(\$104{,}000) = \$13{,}700$$

Subtotal, indirects: $24,500

The total annual cost is the sum of the directs and indirects, or:

$$\text{TAC} = \$32{,}300 + \$24{,}500 = \$56{,}800$$

Notice that over one-third of the TAC is due to the fuel cost alone.

*   *   *

With this we end our three-chapter coverage of the "add-on" controls. Although these "add-ons" and their auxiliary equipment are the mainstays of air pollution control, other kinds of equipment have been devised to control emissions from (if you will) "nontraditional" sources. We'll cover some of these in the next chapter.

REFERENCES

1. Katari, V. S., W. M. Vatavuk, and A. H. Wehe. "Incineration Techniques for Control of Volatile Organic Compound Emissions, Part I: Fundamentals and Process Design Considerations," *Journal of the Air Pollution Control Association*, January 1987 (Vol. 37, No. 1), pp. 91–99.
2. Perry, R. H., C. H. Chilton, and S. D. Kilpatrick, Eds. *Chemical Engineers' Handbook*. New York: McGraw-Hill, 1963, pp. 3–142 to 3–144.
3. *Handbook of Chemistry and Physics*, 54th ed. Boca Raton, FL: CRC Press, Inc., 1973–74, pp. D85–D92.
4. Price and technical data from Salem Industries, Inc. (South Lyon, MI), December 1988.
5. Katari, V. S., W. M. Vatavuk, and A. H. Wehe. "Incineration Techniques for Control of Volatile Organic Compound Emissions, Part II: Capital and Annual Operating Costs," *Journal of the Air Pollution Control Association*, February 1987 (Vol. 37, No. 2), pp. 198–201.
6. Oakes, D. W. "Practical Applications of Solvent Emission Control Using Activated Carbon." Presented at the 75th Annual Meeting of the Air Pollution Control Association (#82–18.6), June 20–25, 1982.
7. *EAB Control Cost Manual,* 3rd ed. Section 4: "Carbon Adsorbers." Research Triangle Park, NC: U.S. Environmental Protection Agency, 1987. (EPA/450/5–87–001A; NTIS PB–87–166583/AS).
8. Price and technical data from Tigg Corporation (Pittsburgh, PA), November 1987.
9. Price and technical data from Hoyt Corporation (Westport, MA), December 1988.
10. Vatavuk, W. M., and R. B. Neveril. "Estimating Costs of Air-Pollution Control Systems, Part XVI: Costs of Refrigeration Systems," *Chemical Engineering*, May 16, 1983, pp. 95–98.
11. Price and technical data from Edwards Engineering Corporation (Pompton Plains, NJ), August 1988.
12. Vatavuk, W. M., and R. B. Neveril. "Estimating Costs of Air-Pollution Control Systems, Part XV: Estimating Costs of Flares," *Chemical Engineering*, February 21, 1983, pp. 89–90.
13. Oenbring, P. R. "Flares and Flare System Design." In: *Encyclopedia of Chemical Processing and Design,* Vol. 22. New York: Marcel Dekker, 1985, pp. 144–157.
14. Price and technical data from Sur-Lite Corporation (Santa Fe Springs, CA), December 1988.

CHAPTER 7

A Potpourri of Control Costs

What is it all but a Woolworth welter of things?
Seven Munich Elegies 5 — George Barker

The last three chapters dealt with sizing and costing the more commonly used "add-on" control devices and their auxiliaries. This chapter addresses other control technologies, some for "stack" type emission sources, others for "nontraditional" sources, such as controlling fugitive particulate from uncovered storage piles. The technologies covered here are:

1. Wet dust suppression
2. Coal cleaning
3. Flue gas desulfurization systems (FGDs)
   - Throwaway
   - Regenerable
4. Soil biofilters

Although none of these technologies are covered in the same level of detail as were the "add-on" devices, enough information will be provided to enable one to make "study" or "order-of-magnitude" cost estimates.

WET DUST SUPPRESSION SYSTEMS (WDS)

Description and Design Considerations

Among the largest sources of fugitive particulate emissions are uncovered storage piles. Strong winds and even light breezes can scatter coal, metal ores, limestone, crushed rock, waste, and other dust over large areas, causing not only an air pollution problem, but in some cases, losses of valuable materials. To control these emissions, *wet dust suppression systems* were devised. Their design and operating principles are quite basic.

A typical WDS consists of an arrangement of spray stations (stanchions) located along the perimeter of a storage pile (or piles). Each station is piped to the others and to a central proportioning and pumping station, which conveys a solution of water and a "surface active" (wetting) agent to it. When sprayed on the pile, the wetting agent agglomerates the dust into larger particles of

Figure 7.1 Wet dust suppression system controlling long-term storage pile (courtesy Johnson-March Systems, Inc.).

greater mass. In turn, this creates a surface which is highly resistant to wind erosion. The solution flowrate to the spray is metered by a controller, usually mounted on the spray stanchion. Finally, the entire system is automated via a remote control panel. (Figure 7.1 depicts a WDS in a typical application.)[1]

Several variables affect the design and operation of a WDS: (1) size of particles in the storage pile; (2) weather conditions; and (3) wettability of the pile material. Regarding particle size, it is clear that the smaller the particle, the more easily the wind will carry it away. The particle emission rate will also depend on such weather conditions as the season, air temperature, cloud cover, and wind velocity.

All of these parameters affect the rate water evaporates from the pile. (Typical evaporation rates from a thoroughly wetted pile range from 0.1 [winter day] to 0.8 mm/hour [summer cloudless day].)[1]

In addition, wind erodability tends to vary inversely with material wettability—the amount of adsorbed water film that surrounds dust particles. This water film causes the particles to bind together, making them less subject to wind forces. However, the pile moisture content is not the only factor to consider. Some materials, such as coal, are *hydrophobic* and require the use of a wetting agent to reduce the water surface tension. Moreover, because these wetting agents remain in place longer than water alone, they reduce the required spraying frequency.[1] (However, surfactants can adversely affect the quality of some piled materials.)

In addition, certain system parameters affect the operating efficiency of a WDS. These are the nozzle size, water line pressure, and spraying angle.

Larger nozzles can deliver more water per unit time but require larger (and more expensive) piping systems. By increasing the line pressure, one can spray solution for greater distances, but at the cost of higher pump horsepower and electricity consumption. Thirdly, by lowering the spray angle, one can increase the "reach" of a spray stanchion. However, this might make it more difficult to reach the tops of higher piles. In a well-designed WDS, these parameters are optimized.[1]

Finally, both the design and operation of a WDS will depend on whether the pile being sprayed is "short term" (active) or "long term" (inactive). To control dust from active piles, the water and wetting agent (surfactant) mixture is sprayed onto the pile often enough to maintain the moisture level of the pile surface. The spraying frequency and duration depend upon the atmospheric conditions (i.e., wet or dry). Between spray periods, as the very top surface of the pile dries, the sub-surface moisture migrates upward to replenish the surface moisture. By reducing the surface tension, the surfactant makes this easier to accomplish.[1] Dust at long-term storage piles is typically controlled by adding a *binding* agent to the pile surface to bind the surface particles together. Although the surface remains porous, because the dust particles are bound to the surface, the wind cannot remove them. Unlike short-term piles, which must be resprayed often, long-term piles so treated will remain dust-free for a year or more. The binding compound is usually applied by a contractor (e.g., a local tree sprayer), with periodic touchup spraying being done by source personnel.

Costing Procedure

Generally, the capital and annual costs of a WDS correlate with the surface area of the pile to be sprayed (in ft²). Also, the costs for WDS spraying active (short-term) piles depend on the *shape* of the pile—i.e., "conical" or "row/flat top row." A vendor supplied the following cost data (all in **December 1988** dollars) for the WDS he manufactures:[1]

Total Capital Investment (TCI)

Costs provided were regressed to fit equations of the form:

$$TCI\ (\$) = aS^b \qquad (7.1)$$

where S = pile surface area (ft²)

Table 7.1   Direct Annual Costs of Wet Dust Suppression Systems[a]

Cost Element	Cost/Quantity (Pile Type)	
	Short-term	Long-term
Maintenance		
Labor (days/yr)	$3A^{0.176}$ [b]	$2A^{0.301}$
Materials ($/yr)	$350A^{0.243}$	$184A^{0.500}$
Water (1000 gal/hr)	8A	None[c]
Surfactant ($/hr)	16.6A	None
Electricity (kW)	70A	None

Source: Johnson-March Systems, Inc. (Ivyland, PA).
[a]All costs are in **December 1988** dollars.
[b]All cost equations are valid for pile level areas (A) from 0.1 to 10.0 acres.
[c]These costs have been included in the capital (contractor's) cost.

Values for these parameters are as follows:

Pile Type/Shape	Correlation Range (Surface Area 1,000 ft²)	Parameter	
		a	b
Short-term			
Row:	8 to 800[a]	3930	0.363
Conical:	4.5 to 45	20700	0.185
Long-term	8 to 800[a]	4.88	0.652

---

[a]For these pile types, the *surface* area is 80,000 ft^2 per acre of *level* area.

The TCI for the short-term pile WDS includes costs for designing the system, the equipment, and all installation charges. However, because the long-term piles are sprayed by contractors, the TCI includes the costs for equipment rental, application labor, and binding compound.

Annual Costs

These costs will vary not only with the pile size and type but also with the system annual operating hours. Values/equations for computing WDS *direct* annual costs are tabulated in Table 7.1. The cost equations in the table apply to storage piles with surface areas ranging from 0.1 to 10.0 acres. Also, note that the costs for the short-term piles apply to *both* conical and row shapes.

The *indirect* annual costs for WDS include overhead, capital recovery, and property taxes, insurance, and administrative charges. As in previous chapters, overhead is 60% of the sum of maintenance labor and materials costs. Property taxes, etc., are figured at the standard 4% of the TCI. Finally, when computing the capital recovery cost, use the applicable annual interest rate and a 25-year system life.[1]

Example: An owner of a stone quarry needs to install a wet dust suppression system to control fugitive emissions from an active gravel pile. The pile is "row"-shaped and occupies a level area of 7.5 acres. If the WDS is to operate for 1000 hr/yr, estimate its capital and annual costs.

Solution: As indicated by the cost correlation parameters above, a row pile has a surface area of 80,000 ft^2 for every acre of level land area. For this pile, the surface area would be:

$$S = 7.5 \times 80,000 = 600,000 \text{ ft}^2$$

Substitution of this area into Equation 7.1 and the Table 7.1 cost equations yields:

• Total capital investment:

$$\text{TCI} = 3,930(600,000)^{0.363} = \$492,000$$

• Direct annual costs:
Maintenance labor $= 3(7.5)^{0.176} = 4.3$ days/yr
At an $11/hr rate, this would be:
Maintenance labor cost $= (4.3)(8)(\$11) = \$380$
Maintenance materials $= 350(7.5)^{0.243} = \$570$
Water $= 8(7.5) = 60,000$ gal/hr
At a \$0.20/1000 gal cost and 1000 hr/yr, we have:

$$\text{Water cost} = \$12,000$$

Surfactant $= 16.6(7.5)$ (\$/hr) $\times$ 1000 hr $= \$124,500$
Electricity $= 70(7.5) = 525$ kW
At a cost of \$0.06/kWhr, we have:

$$\text{Electricity cost} = (525 \text{ kW})(1000 \text{ hr})(\$0.06/\text{kWhr}) = \$31,500$$

Total, direct costs: \$169,000
• Indirect annual costs:
Overhead $= (0.60)(\$380 + \$570) = \$570$
Property taxes, etc. $= (0.04)(\$492,000) = \$19,680$
Capital recovery: Based on a 25-year life and a 10% annual interest rate, we calculate a capital recovery factor (CRF) of: 0.1102. Therefore:

$$\text{CRC} = (0.1102)(\$492,000) = \$54,220$$

Total, indirect costs: \$74,500
Total annual cost $= \$169,000 + \$74,500 = \$243,000$

COAL CLEANING

Description

Coal cleaning (also known as coal beneficiation/preparation) encompasses several processes used to enhance the properties of mined ("raw") coal by reducing the amounts of such impurities as ash and sulfur. Because these impurities are denser, they can be separated from the bulk of the coal via such gravity-based techniques as jigs, concentrating tables, and hydrocyclones. Although the intent of these "physical" cleaning methods has been to remove ash and other noncombustible materials, in recent years they also have been used to reduce the coal *pyritic* sulfur content—the sulfur not chemically bonded to the coal. (The coal *organic* sulfur content can only be reduced by chemical means, however.) Because the removal of pyritic sulfur eventually reduces the quantity of sulfur oxides generated by the coal when it is burned, coal cleaning is often considered more cost-effective than flue gas desulfurization and other SO_x control methods.

The size of the coal to be washed, the desired final properties (particularly, ash and sulfur contents), and the "preparation level" determine the types of coal cleaning processes to be used. For example, at a high preparation level, the coal would first be crushed and screened. Then, the largest ("top") sizes would be processed in a "heavy medium vessel," while the medium and fine sizes would be treated in a "heavy medium cyclone" and hydrocyclone, respectively. In these processes, coal impurities settle out, while the cleaned coal ("float") is skimmed off. As expected, the higher the preparation level, the cleaner the coal—and the higher the cleaning cost.[2]

Costing Procedure

Skea and Rubin[2] developed a comprehensive cost model for coal cleaning plants that correlates the capital and annual costs with such variables as the inlet ("raw") coal mass flowrate, annual operating hours, and the preparation level. We escalated the model's capital costs from 1978 to **June 1988** dollars, via the composite Chemical Engineering Plant Index. However, we escalated the annual costs using the Consumer Price Index (CPI-U). Because of this long escalation period—and because some of their original costs (e.g., the capital cost equation scaling exponent) were selected arbitrarily—the costs shown below should be considered "order-of-magnitude" estimates ($> \pm 30\%$ accurate).

Total Capital Investment

Equation 7.2 below estimates the TCI for *new* coal cleaning plants with capacities of 500 to 2000 tons/hr of clean coal. It includes costs for five plant sections: (a) raw coal handling, (b) coal cleaning equipment, (c) thermal dry-

Table 7.2   Direct Annual Costs of Coal Cleaning Plants[a]

Cost Element	Cost (Preparation Plant Level)		
	2	3	4
Chemicals ($/raw ton)	0.045	0.079	0.12
Electricity ($/raw ton)	0.12	0.24	0.27
Labor ($/hr)	230	240	400
Maintenance (fraction of [TCI − WC])	0.07	0.07	0.07
Refuse disposal ($/wet ton)	1.8	1.8	1.8
Water ($/raw ton)	0.0020	0.0032	0.0035

Source: Skea, J. F., and E. S. Rubin. "Optimization of Coal Beneficiation Plants for SO_2 Emissions Control," *Journal of the Air Pollution Control Association*, October 1988, pp. 1281–1288.
[a]All costs have been escalated to **June 1988** dollars via the Consumer Price Index (CPI-U).

ing, (d) refuse handling, and (e) coal sampling. The costs for the first two items are functions of the raw coal mass flowrate (M_{raw}, dry tons/hr). The thermal drying and refuse handling costs have been correlated against the thermal dryer water evaporation rate (M_{we}, tons/hr) and the coal refuse mass flowrate (M_{ref}, dry tons/hr), in turn. Finally, the portion of the TCI attributable to the coal sampling section is assumed to be constant for all plant sizes.[2]

$$\text{TCI (1000 \$)} = 770 + (31 + k)M_{raw}^{0.7} + 310M_{we}^{0.7} + 75M_{ref}^{0.7} + WC \qquad (7.2)$$

where   WC = working capital
        = 0.25 × direct annual costs

The value of "k" in this equation depends on the level of preparation, as follows: level 2 − 50; level 3 − 86; level 4 − 110.

Direct Annual Costs

These consist of costs for chemicals, electricity, labor, maintenance, water, and refuse disposal. These costs are tabulated in Table 7.2. Notice that except for labor and maintenance, the costs are expressed on a $/ton basis. Also note that the maintenance factor (7%) is applied against the TCI *minus* WC (the working capital). Like the TCI, these costs have been escalated to **June 1988** dollars.

Indirect Annual Costs

These include the usual items:

- Overhead: 60% of the sum of operating and maintenance labor, plus maintenance materials
- Property taxes, insurance, administrative charges:

4% of (TCI − WC)

- Capital recovery, computed as:

$$CRC = (CRF)(TCI - WC) + (WC)(i) \qquad (7.3)$$

where   CRF = capital recovery factor for the plant life and interest rate in question

Notice that the capital recovery calculation is made in two steps. In the first, the CRF is applied against the TCI less the WC, because the working capital is *not* a depreciable expenditure. Instead, WC is treated as if were a sum of money "held" for the life of the plant and then "returned." During that period, however, one would have to pay (simple) interest on that sum. The amount of interest would just be the product of the interest rate and the working capital. (For a fuller discussion of working capital, see Chapter 2.)

FLUE GAS DESULFURIZATION SYSTEMS

The combustion of fossil fuels (mainly coal), by such sources as steam-electric power plants and industrial boilers, is by far the largest human-generated source of sulfur dioxide (SO_2) emissions to the atmosphere. Traditionally, two approaches have been taken to reduce these emissions: (1) reduce the sulfur fed to the combustor, either by burning fuels with low sulfur contents and/or by removing sulfur from the fuel beforehand (e.g., via coal cleaning) and (2) removing SO_2 from the combustor exhaust. The latter approach typically involves the use of *flue gas desulfurization systems* (FGDs). Nearly twenty different types of FGDs have been developed, each of which removes SO_2 from the flue gas via an absorption process. For convenience, FGDs are classified either as "throwaway" or "regenerable," depending on whether the absorber product is treated to recover the reagents or simply disposed of. FGDs are further subdivided according to the sulfur content of the coal being burned. In this section, we'll briefly describe these FGDs and present average capital and annual costs (in **June 1988** dollars) for them. With each description, we'll indicate the level of development for the process by a letter, viz.: A—commercial; B—full-size (i.e., demonstration project–scale); C—pilot plant; and D—bench-scale.

"Throwaway" FGD Systems

The majority of the FGDs are "throwaway" processes, most of which have been (or can be) used to scrub high-sulfur coal flue gases at steam-electric power plants. These include the following 10 processes:[3]

1. *Conventional Limestone (A)*: In this process, a limestone slurry solution is injected in a spray tower to absorb SO_2 and form a calcium

sulfite/sulfate sludge. Limestone scrubbing is widely applied commercially and utilizes a cheap, abundant absorbent. However, because it can remove up to 90% of the inlet SO_2 and is quite reliable, limestone scrubbing is often the "FGD of choice." Disadvantages: it's subject to scaling, plugging, erosion, and corrosion, mainly due to the circulating alkali slurry.

2. *Chiyoda Thoroughbred (C-T) 121 (B)*: This process also uses a limestone slurry for SO_2 removal, but does so in a single vessel jet bubbling reactor, not a conventional absorber. Calcium sulfate is the end product. However, the C-T 121 process uses the limestone slurry more efficiently and doesn't suffer with the scaling, corrosion, and other operating problems that plague limestone FGDs. Also, because slurry pumps are not required, the C-T capital cost is lower.

3. *DOWA (C)*: In this process, an aluminum sulfate solution absorbs the SO_2 and is oxidized and then regenerated with limestone to form a gypsum product. Relative to conventional limestone FGD, the DOWA process waste has superior structural properties and is more filterable and settle-able. Other advantages include more efficient limestone utilization and better turndown, though these may be moderated by local restrictions on disposal of the aluminum compounds in the waste.

4. *Forced Limestone Oxidation (FLO) (A)*: As in conventional limestone, SO_2 is absorbed by a limestone slurry in a spray tower. But following this, the waste is oxidized to form a gypsum product. FLO has the same advantages as the DOWA process, relative to the waste properties.

5. *Saarberg-Holter (A)*: This Swedish process employs a clear lime solution to absorb sulfur dioxide in a "Rotopart" scrubber and then to form a calcium sulfite/sulfate precipitate that is oxidized in the same scrubber to form a gypsum product. In addition to offering a more efficient reagent utilization and superior waste properties, this process has better turndown and load-following capabilities.

The Saarberg-Holter process is similar to the C-T 121 and FLO processes, in that all three use single-loop systems to produce gypsum. The S-H and C-T 121 capital costs are approximately equal, although the former's variable operating costs are higher than the others' because the S-H uses lime, a more expensive reagent than limestone.

6. *Limestone Dual Alkali (B)*: Here, SO_2 is removed by a NaOH solution in a spray tower. Limestone added to the product solution acts both to regenerate the sodium solution and to form a calcium sulfite/sulfate sludge. Relative to the conventional limestone process, this FGD offers lower scaling, plugging, corrosion, and erosion potential. Because the sodium reagent is more reactive than the calcium compounds used in the above FGDs, a lower liquid/gas (L/G) ratio is tolerated, translating to lower pumping power requirements and better turndown and load-following capabilities. But because the waste contains soluble sodium salts, a more elaborate waste disposal site may be needed.

7. *Lime Dual Alkali (A)*: Like the limestone dual alkali and DOWA systems, this is a dual-loop process. In the first loop, a clear sodium sulfite solution absorbs SO_2 in a spray tower. Lime is added to the product solution in the second loop external to the scrubber, wherein the sodium solution is regenerated and a sludge is formed. Lime Dual Alkali offers the same advantages as Limestone DA, with respect to the lower L/G ratio, pumping power consumption, and capital cost. However, its variable operating cost is the highest of all because it uses the more expensive lime, instead of limestone.

8. *Wet (Conventional) Lime (A)*: This is the counterpart to the conventional limestone FGD process. Although its reagent cost (lime) is higher, it suffers less erosion, nearly total reagent use, and better turndown and load-following capabilities. Moreover, to prepare the lime slurry, raw water is required, instead of the less expensive cooling tower blowdown water used with limestone-based systems.

9. *Lime Spray Dryer (A)*: This is a semi-dry process in which the flue gas and a lime slurry mix in a spray dryer. The flue gas SO_2 and lime react to form a solid, which is collected with the fly ash in a fabric filter immediately downstream. Its capital and operating costs and maintenance and energy requirements are lower than those for conventional limestone scrubbing, and its waste handling is less bothersome. On the down side, the filter bags can blind if the flue gas approaches saturation temperature, and scaling can occur in the spray dryer.

10. *Nahcolite or Trona Injection (B)*: Unlike the Lime Spray Dryer process, this technique involves injection of a dry reagent (nahcolite or trona) into the flue gas duct to react with the SO_2. The waste is then captured in a downstream baghouse. Because this process employs a totally dry waste, less process equipment is required, and as a result, capital and operating costs, and maintenance, energy, and waste handling requirements are lower. However, the reagent is hard to

obtain and is less efficient below 275°F; also, the waste it generates may require more elaborate (and more costly) disposal methods.

Trona has been found to be a less expensive and easier-to-obtain substitute for nahcolite as a dry injectant. But Trona Injection is not as well-developed as the Nahcolite process.

Regenerable FGD Systems

Although regenerable FGDs generally have higher capital costs than "throw-away" processes, they are advantageous when space or waste disposal requirements are limited and/or when there are readily available markets for the system by-products.

1. *Wellman-Lord (W-L) (A)*: The Wellman-Lord has been used in flue gas desulfurization applications more than any other regenerable process. First, a sodium sulfite solution contacts the SO_2 in a tray tower to produce a rich sodium bisulfite solution. The latter is regenerated to sodium sulfite in a steam evaporator. The SO_2 gas stream released during the regeneration is converted to sulfur in an Allied Chemical plant onsite. The W-L process produces a salable by-product (sulfur) and is free of scaling and plugging problems. However, working against this are its high operating costs, its need for an expensive purge system, and its sensitivity to particulates, HCl, and SO_3 (sulfur trioxide).

2. *Magnesium Oxide (A)*: In the "Mag-Ox" system, a magnesium oxide solution removes SO_2 in a grid-packed spray tower and forms a magnesium sulfite/sulfate product. The solids are calcined to release SO_2 gas and reusable magnesium oxide. The SO_2 is processed, in turn, into sulfuric acid, which can be sold. Compared to the Wellman-Lord process, which consumes steam, the Mag-Ox produces it. But there are several disadvantages, such as a very high oxidation rate and potential equipment pluggage and erosion.

3. *Sulf-X (C)*: In this pilot-scale process, an iron sulfide slurry solution removes sulfur dioxide in a packed tower. Regeneration of the product in a calciner produces an elemental sulfur offgas, which is condensed to a salable liquid form. Its low oxidation rate and fuel consumption, coupled with NO_x removal capabilities, are somewhat offset by potentials for equipment erosion, process control problems, and other disadvantages.

4. *Flakt-Boliden (C)*: Sulfur dioxide is removed in a packed tower by a sodium citrate solution. Following solution regeneration in a stripping column, the concentrated SO_2 stream is converted to elemental sulfur in another Allied Chemical plant. Like the Sulf-X, the Flakt-Boliden has a

lower oxidation rate than the Wellman-Lord. However, the vacuum crystallization process used in the sulfate-purge system is harder to control than the thermal crystallization system in the Wellman-Lord process.

5. *Aqueous Carbonate (C)*: Absorption of SO_2 by a sodium carbonate solution in a spray dryer is followed by regeneration of the solution in a molten-salt bed using coke. The H_2S (hydrogen sulfide) stream generated therein is converted to sulfur in a Claus plant. Although this process consumes less steam and is less sensitive to in-absorber oxidation than the W-L, it is extremely complex and, consequently, is potentially unstable and prone to equipment failures.

6. *Conosox (C)*: This process uses a mixture of potassium carbonate and potassium salts to remove SO_2 in a packed tower, forming a potassium bisulfite product. After being converted to a potassium thiosulfate solution, this product is reduced with carbon monoxide to form regenerated potassium carbonate and H_2S gas. The latter is reduced to sulfur in a Claus plant. Compared to the W-L, the Conosox uses a more reactive reagent and has lower oxidation and steam consumption rates, while having a lower absorber pressure drop. But it has several disadvantages, such as the need for high process temperatures and pressures and high fuel and liquid oxygen consumptions.

Costs for These FGD Processes

The Electric Power Research Institute (EPRI) has developed capital and annual costs for each of the above 16 FGD processes. Each FGD has been sized to control two 500-megawatt coal-fired power plant boilers located in Wisconsin. The FGDs operate 5694 hr/yr, corresponding to a 65% "capacity factor." Originally in December 1982 dollars, the EPRI costs have been escalated to a **June 1988** reference date using the Chemical Engineering Plant index. Because FGDs have most commonly been applied to the control of steam-electric power plants, the costing procedures and formats used in developing the EPRI costs are those traditionally used by the utility industry. Accordingly, the capital costs are given on a "$/kW of generating capacity" basis, while the annual costs are expressed in terms of "mills/kWh" – the so-called "levelized busbar cost" (LBC). The LBC is the utility industry's preferred way to calculate the total annual cost (TAC). However – and we can't emphasize this enough – the LBC and TAC calculation procedures are NOT equivalent. As Chapter 2 indicates, the TAC method rests on a constant-dollar analysis (i.e., O&M costs and capital-related charges are assumed *not* to increase due to inflation during the lifetime of the project). The LBC method does factor in inflation in computing the "levelized" cost of a project over its lifetime – typically, 30 years for a FGD system. For more information on the LBC method, consult reference 4.

Table 7.3   Capital and Levelized Busbar Costs for Throwaway FGD Processes[a]

Process	Cost	
	Capital ($/kW)	Levelized Busbar (mills/kwhr)
Limestone[b]	120–190	10–22
Chiyoda Thoroughbred 121	150	17
DOWA	190	17
Forced limestone oxidation	190	19
Saarberg-Holter	140	19
Limestone dual alkali	170	19
Lime dual alkali	160	21
Wet lime	180	24
Lime spray dryer	120	9
Nahcolite/trona injection	27	8–10[c]

Source: Keith, R. J., J. E. Miranda, J. B. Reisdorf, and R. W. Scheck. *Economic Evaluation of FGD Systems, Volume 1: Throwaway FGD Processes, High- and Low-Sulfur Coal* (CS-3342). Palo Alto, CA: Electric Power Research Institute, December 1983.
[a]Capital costs have been escalated to **June 1988** dollars via the Chemical Engineering Plant Cost Index (composite). Levelized busbar (annual) costs have been escalated via the Consumer Price Index (CPI-U).
[b]Lower and higher costs correspond to control of two 500-MW units burning low- and high-sulfur coals, respectively. Units operate 5,694 hr/yr.
[c]Lower cost pertains to trona injection; higher cost, to nahcolite injection.

Tables 7.3 and 7.4 show capital costs and LBCs for the throwaway and regenerable FGDs described above, respectively.[3] Recall that these costs pertain to controlling two 500-MW boilers, each operating 5694 hr/yr. Clearly, the costs would be different for boilers of different sizes. Further, such variables as the coal sulfur content, geographic location, and local costs for labor, reagent, and waste disposal also would affect these expenditures. If the Table 7.3 and 7.4 data were used to calculate costs for FGDs of different sizes, operating hours, etc., the resulting estimates would only be of order-of-magnitude accuracy ($> \pm 30\%$).

Table 7.4   Captial and Levelized Busbar Costs for Regenerable FGD Processes[a,b]

Process	Cost	
	Capital ($/kW)	Levelized Busbar (mills/kwhr)
Wellman-Lord	300	31
MGO	290	23
Sulf-X	320	24
Flakt-Boliden	420	35
Aqueous Carbonate	430	36
Conosox	460	54

Source: Keith, R. J., J. E. Miranda, J. B. Reisdorf, and R. W. Scheck. *Economic Evaluation of FGD Systems, Volume 1: Throwaway FGD Processes, High- and Low-Sulfur Coal* (CS-3342). Palo Alto, CA: Electric Power Research Institute, December 1983.
[a]Capital costs have been escalated to **June 1988** dollars via the Chemical Engineering Plant Cost Index (composite). Levelized busbar (annual) costs have been escalated via the Consumer Price Index (CPI-U).
[b]Each of these FGDs has been sized to control two 500-MW boilers firing high-sulfur coal and operating 5,694 hr/yr.

SOIL BIOFILTERS

Among the more unique (and ingenious) control methods to be applied in recent years are soil biofilters — soil beds, for short. These systems have proved to be very effective in first adsorbing, then (if possible) oxidizing a range of pollutants: volatile organic compounds, SO_2, NO_x, and H_2S. Control efficiencies have ranged from 90 to 99 + % for these and other compounds, under typical soil conditions. Soil beds have been used to control waste gases at installations in the chemicals, pharmaceuticals, and food processing industries.[5] A soil bed consists of a horizontal network of perforated pipe installed about 2 ft below the surface. The network uniformly disperses an air-waste gas stream throughout the bed. This bed functions as an adsorption system, capturing the waste gas molecules on soil particles, while air, water vapor, and carbon dioxide pass upward through the soil pores. The adsorbed molecules are oxidized, either microbiologically or via surface catalysis. The oxidation both converts the pollutants to harmless H_2O and CO_2 and regenerates the soil bed. The inherent biodegradability of the gases and the rate of microbial activity determine this oxidation rate. The latter depends on the temperature (100°F is optimal) and the supplies of food, oxygen, water, and other nutrients.[5]

Along with removing organic gases, soil beds can also adsorb inorganics (e.g., SO_2) and particulates. However, because the latter can plug the soil pores, eventually rendering them ineffective, they should be removed from the waste gas upstream of the bed.

Finally, the costs of soil biofilters are relatively modest. The capital investment — for assembling and installing an underground pipe network for gas distribution — is approximately $8 to $10/acfm of gas stream treated. The only direct annual cost is electricity for powering a fan to overcome the 2- to 3-in. w.c. pressure drop through the bed. This corresponds to a power consumption of 0.4 to 0.6 watts/acfm. No labor, maintenance, or other O&M costs would be required.[5]

Finally, because a soil biofilter has no parts requiring replacement, it should last a long time — at least 20 years. At a 10% annual interest rate, this would translate to capital charges (capital recovery, property taxes, etc.) of roughly $1 to $2/acfm-yr.

REFERENCES

1. Price and technical data from Johnson-March Systems, Inc. (Ivyland, PA), December 1988.

2. Skea, J. F., and E. S. Rubin. "Optimization of Coal Beneficiation Plants for SO_2 Emissions Control," *Journal of the Air Pollution Control Association*, October 1988, pp. 1281–1288.

3. Keeth, R. J., J. E. Miranda, J. B. Reisdorf, and R. W. Scheck. *Economic Evaluation of FGD Systems, Volume 1: Throwaway FGD Processes, High- and Low-*

Sulfur Coal (CS-3342). Palo Alto, CA: Electric Power Research Institute, December 1983.

4. Keeth, R. J., J. E. Miranda, J. B. Reisdorf, and R. W. Scheck. *Economic Evaluation of FGD Systems, Volume 3: Appendixes* (CS-3342). Palo Alto, CA: Electric Power Research Institute, December 1983.

5. Bohn, H., and R. Bohn. "Soil Beds Weed Out Pollutants," *Chemical Engineering*, April 25, 1988, pp. 73–76.

CHAPTER 8

Escalating Costs

You can never plan the future by the past.

Edmund Burke

As you've probably noticed by now, this book contains a lot of cost data. Some of these costs are relatively recent (one year or newer) and were included here "as is." Other costs are not so new—some, in fact, are several years old. To be of any use to the book and its readers, these older costs have had to be adjusted—"escalated"—to a more recent time. The date "June 1988" appears quite frequently herein. We chose June 1988 because it was the latest date for which we could obtain complete cost *escalation index* data. By the time you read these lines, more current index information will have been published, and the costs in this book will be somewhat dated. However, by using the techniques and index data in this chapter, *most* cost data can be given a face lift.

ESCALATION INDICES: WHAT, WHY, AND HOW?

One might ask—and rightfully so—"Why use indices at all? Why not simply obtain more current control cost data from vendors, the literature, or other reputable sources?" For one thing, obtaining current costs is much easier said than done. This is especially true of equipment costs.

As we explained in Chapter 3, the most reputable sources of control equipment costs are the vendors of that equipment. A few vendors, in the spirit of cooperation (or perhaps to obtain some free publicity for their firms) will provide budget quotes for their equipment—as long as the information request is reasonably specific. But most would rather not be bothered. The time and effort to prepare quotes that would never lead to an equipment order can be spent in better ways. And for this attitude they can't be blamed.

On the other hand, obtaining current data for O&M costs—labor, electricity, etc.—is usually less bothersome—as long as the estimator has the time and patience to do some digging. Costs for these items are periodically compiled by such entities as the Bureau of Labor Statistics and the Energy Information Administration, arms of the Departments of Labor and of Energy, respectively. With current price data so readily available, it makes little sense to compile indices for O&M costs. Hence, in air pollution control costing, indices are almost always used to escalate equipment costs only. (One notable exception is aggregated capital and annual control costs, such as those appearing in Chapter 1. As we'll later show, these costs are escalated differently from equipment costs.)

How are escalation indices used? The following equation illustrates:

$$\text{Cost}_2 = \text{Cost}_1(\text{Index}_2/\text{Index}_1) \qquad (8.1)$$

In this equation, the subscript 1 refers to the "base" date—the date *from* which the equipment cost (Cost_1) is being escalated. "Index_1" denotes the value of the index at the base date. Similarly, subscript 2 corresponds to the "reference" date—the date *to* which Cost_1 is to be escalated.

As a rule, "generic" indices (e.g., the Chemical Engineering Plant) should NOT be used to escalate equipment costs over periods exceeding *five years*. That is because no generic index exactly "tracks" price changes for any given type of equipment. Over short periods, the differences between the actual prices and those predicted by a generic escalation index will be small relative to the inherent error in most vendor "budget quotes" ($\pm 20\%$). But over longer periods, these deviations can become significant. (However, even over short periods, indices may be of little value for predicting equipment costs. This is especially true during times of extremely sharp price increases, such as the recent surge in stainless steel prices.)

The five-year limit may not apply to indices based on cost data for a particular industry (such as the Nelson-Farrar Refinery Indexes). An industry-specific index is usually based on large amounts of data collected over long periods of time from many (if not most) firms in the industry. Because the index rests on a very broad, solid foundation, it is extremely representative of industry cost trends. Needless to say, for the index to be reliable, it must be used to escalate costs for that industry only. Though indices may have their drawbacks, they are certainly easy to use. As Equation 8.1 shows, to escalate a cost from one period to another, all one has to do is multiply the cost by the ratio of the two index values for the dates in question. The index values are usually three- or four-digit numbers (e.g., 379.5) that are tied to some base date, at which time the value of the index was 100 (e.g., 1957–59). Occasionally, the composition of the index is revised by the compilers, who may also take this occasion to change the base date. Revisions to indices create discontinuities in the index-time function. However, if the changes are carefully selected, the transition can be made smoothly.

SOME USEFUL COST INDICES

For escalating the prices of air pollution control equipment, NO published index is ideal. The variety of equipment comprising control systems—vessels, pumps, fans, ductwork—is so great that no index can accurately track their prices. Nonetheless, three published indices are designed so that they can be used in escalating control equipment. These are the *Chemical Engineering Plant Cost Index* (CEP), the *Marshall & Swift Equipment Cost Index* (M&S), and the *Producer Price Index*. Following is a description of each.

Chemical Engineering Plant Cost Index

This index, updated monthly, is published in the "Economic Indicators" department of *Chemical Engineering* magazine, a McGraw-Hill publication. Established in 1963, it has been intended for use in escalating chemical process plant construction costs. The index consists of a composite ("CE Index") and 66 components, each weighted according to its contribution to the composite. The composite is often used to escalate total process plant costs, while the components are applied to updating the costs of individual items, both equipment and labor. The component groups and the weights given them are:[1]

- Equipment (61%)
  Heat exchangers and tanks (37%)
  Process machinery (14%)
  Pipe, valves, and fittings (20%)
  Process instruments (7%)
  Pumps and compressors (7%)
  Electrical equipment (5%)
  Structural supports and miscellaneous (10%)
- Construction labor (22%)
- Buildings (7%)
- Engineering and supervision (10%)

In calculating the value for the index, prices for these components are first obtained from the Bureau of Labor Statistics' (BLS) Producer Price Index (PPI). (The BLS obtains its data from widespread surveys of businesses and manufacturers throughout the country, in turn.) Then, the price for each component is multiplied by its weighting factor, above. (The weighting factors have been obtained from an extensive survey of chemical processing firms.) The sum of these price-weighting factor products becomes the composite CE Index for that month. Note that few of these components or subcomponents explicitly correspond to traditional air pollution control system equipment. Nonetheless, the CE Index—specifically, its "Equipment" component—can be used to escalate control equipment costs. (In fact, where escalation was needed in this book, we used this component in nearly all cases.) This component does include a mix of equipment that at least partly corresponds to control system paraphernalia. Moreover, the CE Index is updated monthly, so that it can reflect short-term price changes. (However, the last date for which a "final" index is reported often lags behind the calendar by two to three months.)

Finally, the CE Index was last revised in January 1982. Although the base was not changed at this time (i.e., 1957–59 still equals 100), several of the weighting factors were changed. (The revised factors are shown above). In addition, the "CE productivity factor" was reduced from 2.5 to 1.75% per year. This factor is a "technological productivity factor" that is used to correct the labor components of the subindices for certain advances in working tools and techniques (e.g., proliferation of personal computers). (However, the productivity factor does *not* account for improvements in the productivity of workers or managers.) Table 8.1 lists the CE Index composite and the "Equipment" component from January 1984 to the latest date available.[2]

Marshall and Swift Equipment Cost Index

This index, also published in *Chemical Engineering*, is updated quarterly. Originally known as the "Marshall and Stevens" Index when established in 1937, the M&S Index compiles separate cost indices for 47 commercial, industrial, and housing activities. Reported for the M&S is a composite and two components, each of which is subdivided into several subcomponents:[3]

- Process industries, average
  Cement
  Chemical
  Clay products
  Glass
  Paint
  Paper
  Petroleum products
  Rubber
- Related industries
  Electrical power
  Mining, milling
  Refrigerating
  Steam power

The M&S composite is just the arithmetic average of the 47 individual indices, while "Process industries, average" is the mean of eight selected chemical process industries. Five components comprise each of the industrial indices: (1) process/operating machinery (including tankage, pipes, and fittings), (2) installation labor, (3) power equipment, (4) maintenance equipment, and (5) administrative equipment. These components ("category indices") are assigned weights and used to calculate the individual indices, much as the CE Index components are employed to compute the CE composite. However, each industry index thus calculated is further refined by applying a "bargaining factor" (to reflect normal cost variances, due to location, etc.) and a "premium factor" (which considers abnormal cost differences). These factors can change the unadjusted index value by up to 25%.[3]

Table 8.1   Chemical Engineering Plant Index: 1984 to the Present

Year	Month	CE Index Value	
		Composite	"Equipment" Component
1984:	January	320.3	340.0
	February	320.4	340.4
	March	321.3	342.0
	April	321.9	343.4
	May	322.7	344.1
	June	322.5	344.5
	July	324.1	345.3
	August	323.6	345.5
	September	324.5	345.5
	October	324.1	345.2
	November	323.6	345.9
	December	324.3	346.0
	Annual average:	322.7	344.0
1985:	January	324.7	346.5
	February	325.4	346.8
	March	324.8	346.9
	April	325.5	347.6
	May	325.4	347.0
	June	324.8	347.0
	July	325.2	347.2
	August	325.0	346.7
	September	326.1	347.2
	October	325.8	347.5
	November	325.2	347.6
	December	326.1	348.1
	Annual average:	325.3	347.2
1986:	January	323.5	345.3
	February	319.0	338.1
	March	317.4	336.9
	April	316.7	334.4
	May	317.0	334.2
	June	316.2	333.4
	July	316.9	334.6
	August	317.4	334.6
	September	319.3	336.6
	October	319.3	335.8
	November	318.7	335.6
	December	319.2	335.7
	Annual average:	318.4	336.3
1987:	January	318.3	336.0
	February	318.4	336.9
	March	318.7	337.8
	April	320.1	338.3
	May	321.5	340.0
	June	321.9	340.4
	July	323.9	343.9
	August	324.3	344.9
	September	325.2	345.2
	October	329.8	352.2
	November	330.7	353.8
	December	332.5	357.2
	Annual average:	323.8	343.9

Table 8.1, continued

Year	Month	CE Index Value	
		Composite	"Equipment" Component
1988:	January	336.3	362.8
	February	336.1	363.7
	March	336.5	364.0
	April	340.1	369.4
	May	340.0	369.0
	June	341.6	371.6
	July	343.0	374.2
	August	345.3	376.3
	September	346.2	377.3

Sources: 1. Matley, J., and A. Hick. "Cost Indexes End 1987 on an Upswing," *Chemical Engineering*, April 11, 1988, pp. 71–73. 2. "Economic Indicators," *Chemical Engineering*. (Various issues in 1988.)

Can we use the M&S Index to escalate control equipment costs? At first glance, "No," because the M&S is industry cost–based, while the CE Index is oriented toward generic equipment costs. However, the variety of equipment costs reflected by the 47 industries making up the M&S composite could, in a way, mirror the variety of control equipment used, not only in the chemical process industries, but in other sources (e.g, steam-electric power plants) as well. Further, the M&S and CE composite indices track well—at least in the long run. As reference 4 shows, between 1975 and 1987 (inclusive), the CE Index increased by approximately 77%, while the M&S grew by 82%. But as we'll see shortly, during the last year, steeper equipment price increases have resulted in the two indices going their separate ways.

Table 8.2 lists the M&S composite and "process industries, average" component for the years 1984 to the latest available date. Compare these data with the CE Index data (Table 8.1). For example, between 4th quarter 1987 (November) and 3rd quarter 1988 (August), the M&S composite index increased by 3.6% (856.5/827.0). During that same period, the CE Index composite increased by 4.4% (345.3/330.7). More relevantly, the CE "Equipment" component grew by a whopping 6.4% (376.3/353.8). Those are significant short-run differences. Clearly, the M&S and CE indices cannot be used interchangeably.

Producer Price Index

The Producer Price Index, calculated and published monthly by the Bureau of Labor Statistics, measures average changes in prices received by domestic commodity producers at all "stages of processing." Nearly 7000 individual indices are calculated, based on price reports supplied by producing firms and reports issued by the federal government or various commodity exchanges. For most PPI indices, the base date is 1982 (= 100), changed from 1967 in January 1988.[5]

The individual indices are aggregated in three different ways: (1) *stage-of-*

Table 8.2   Marshall and Swift Equipment Cost Index: 1984 to the Present

Year	Quarter	M&S Index Value	
		Composite	"Process Industries"
1984:	First	770.8	796.4
	Second	781.7	806.8
	Third	783.9	811.2
	Fourth	785.2	811.4
	Annual average:	780.4	806.5
1985:	First	786.6	811.7
	Second	788.7	812.9
	Third	791.1	814.2
	Fourth	792.0	814.9
	Annual average:	789.6	813.4
1986:	First	793.5	815.1
	Second	797.8	817.4
	Third	798.3	816.5
	Fourth	801.0	818.7
	Annual average:	797.7	816.9
1987:	First	803.7	821.5
	Second	808.8	825.1
	Third	814.8	831.3
	Fourth	827.0	843.6
	Annual average:	813.6	830.4
1988:	First	835.3	851.4
	Second	846.7	864.8
	Third	856.5	875.0

Sources: 1. Matley, J., and A. Hick. "Cost Indexes End 1987 on an Upswing," *Chemical Engineering*, April 11, 1988, pp. 71–73. 2. "Economic Indicators," *Chemical Engineering*. (Various issues in 1988.)

processing (i.e, "finished," "intermediate," or "crude"), (2) *commodity classification system*, and (3) *Standard Industrial Classification* (SIC). Of these, the commodity classification system seems best suited to escalating control equipment costs. This system includes 15 "major groups" that contain commodities of similar material composition and end use. A strict coding hierarchy, comprised of separate parallel levels, identifies the groupings. For example:

- 10: Metals and metal products — *Major group*
- 10-7: Fabricated structural metal products — *Subgroup*
- 10-72: Metal tanks — *Product class*
- 10-72-0148: Petroleum storage tanks — *Subproduct class/individual item*

The Producer Price Indices are used by businesses in writing the escalation clauses for long-term contracts, which specify that prices for a given commodity, group of commodities, etc., will increase in accordance with the relevant PPI. In fact, at least one control equipment vendor has used the "Metals and metal products" index as the basis of its price adjustment policy (for wet scrubbers).[6] Therefore, for using the PPI in cost escalation there is ample precedence.

The PPI indices are also invaluable for escalating prices for control equipment components. Consider the bags in a fabric filter. As shown in Chapter 5, the bags can contribute much (if not most) of the total equipment cost. Therefore, if baghouse costs have to be escalated, it seems reasonable to escalate the bag prices using some fabric-related index, while escalating the rest of the baghouse via another index. In fact, we used a PPI index ("Finished Fabrics") to calculate the bag prices for Tables 5.6 and 5.7.

PPI values are shown in Table 8.3 for the "Metals and metal products" major group and the "Finished fabrics" subgroup.

Other Indices in Brief

No discussion of cost indices would be complete without some mention of the *Engineering News Record* (ENR) and the *Nelson-Farrar* indices. Although neither is particularly useful for escalating control equipment costs, each is venerable, well-known, and frequently used by cost estimators.

The ENR Index, established in 1921, is computed biweekly and published in *Engineering News Record* magazine. The ENR is a weighted, aggregative index with three fixed bases (1913, 1926, or 1949 = 100). It is comprised of selected quantities of construction materials (steel, lumber, and cement) and common labor, each multiplied by prices in effect at the time the index is calculated. The materials have been carefully chosen to reflect those used in past, present, and anticipated future construction projects. Common labor has been selected as the labor factor, because wage trends in this category have been found to reflect pay scale changes for skilled, as well as unskilled, construction workers.[7]

The ENR has had a number of applications, such as the assessment of taxes, timing of work starts, fire insurance coverage determinations, and, of course, a wide range of cost estimates. However, because of its composition, the ENR is much more suited to escalating the costs of large construction projects than equipment prices.

The Nelson-Farrar Indices (formerly the "Nelson Refinery Cost Indices") also have many applications. These include adjusting wages and salaries, updating pension plans and payments, calculating updated refinery construction costs, predicting refinery costs, and tracking fuel costs, refinery replacement costs, and refinery operating costs. Calculated and compiled monthly in the *Oil and Gas Journal* since 1946, the Nelson-Farrar Indices track the costs of a long list of petroleum refining–related operating and equipment costs. The operating cost indices include those for electric power, fuel, inorganic and organic chemicals, and labor. Included under the "Equipment or materials" category are cost indices for equipment ranging from "bubble trays" to "valves and fittings." In addition, there are Nelson-Farrar indices that cover construction labor cost, "Refinery Operation," and "Refinery Process Operation." The base dates for these indices are either 1946 or 1956 (= 100).[8]

Although some of the Nelson-Farrar equipment indices could be used to

Table 8.3   Selected Producer Price Indices: 1984 to Present

		PPI Index Value	
Year	Month	"Metal Products"	"Finished Fabrics"
1984:	January	103.7	100.2
	February	104.4	101.4
	March	105.0	101.9
	April	105.4	101.9
	May	105.2	102.2
	June	105.2	102.0
	July	104.8	101.9
	August	104.8	102.1
	September	104.7	102.2
	October	104.8	101.8
	November	104.9	101.7
	December	104.6	101.6
	Annual average:	104.8	101.7
1985:	January	104.4	101.6
	February	104.6	102.0
	March	105.0	102.0
	April	105.0	102.1
	May	104.9	101.0
	June	104.4	100.9
	July	104.3	100.8
	August	104.3	101.1
	September	104.2	101.3
	October	104.2	101.3
	November	103.9	101.2
	December	103.9	101.1
	Annual average:	104.4	101.4
1986:	January	103.1	101.3
	February	103.2	101.1
	March	103.2	101.3
	April	103.1	101.3
	May	103.0	101.5
	June	103.0	101.5
	July	102.9	101.6
	August	103.1	101.6
	September	103.3	101.4
	October	103.4	101.3
	November	103.4	101.3
	December	103.3	101.3
	Annual average:	103.2	101.4
1987:	January	103.7	102.7
	February	103.8	103.0
	March	104.0	103.1
	April	104.4	103.4
	May	105.2	103.7
	June	105.8	104.0
	July	106.7	104.0
	August	107.7	104.6
	September	108.8	104.7
	October	110.5	105.5
	November	111.4	105.9
	December	112.9	106.2
	Annual average:	107.1	104.2

Table 8.3, continued

| Year | Month | PPI Index Value | |
		"Metal Products"	"Finished Fabrics"
1988:	January	114.4	106.9
	February	114.7	108.2
	March	115.4	109.0
	April	116.9	109.2
	May	117.4	109.3
	June	118.0	109.6
	July	119.2	109.4
	August	119.7	109.3
	September	120.3	109.8

Source: U.S. Department of Labor, Bureau of Labor Statistics (Washington, DC), 1988.

escalate control equipment costs (e.g., "Pumps, compressors, etc."), because the indices are geared exclusively to petroleum refineries their application to generic air pollution control systems would be limited at best.

EQUIPMENT COST INDICES COMPARED

Tables 8.1 to 8.3 show values for the three equipment cost indices for the past five years. All well and good. But historically, how well have these indices compared to ("tracked") each other? For instance, which of the three indices has increased the most since 1984? Table 8.4 answers that and similar questions. As the data there show, the Producer Price Index ("Metals and metal products" major group) has increased the most—14.2%—while the Marshall and Swift Index (composite) has grown the least, only 8.5%. The Chemical Engineering Index ("Equipment" component) 1984–88 increase falls between the two, closer to the M&S than the PPI.

To be precise, we should say "changes" in the indices, not increases. For between 1985 and 1986, the CE *decreased* 3.1%. During this year, the cost of living rose only slightly (1.9%, as measured by the Consumer Price Index[9]). More relevantly, the process equipment market was in a slump due to a slow-

Table 8.4 Changes in Equipment Cost Indices: 1984 to Present

| Period | Change (%)[a] | | |
	CE ("Equipment")	M&S	PPI ("Metal Products")
1984–85	0.9	1.2	−0.4
1985–86	−3.1	1.1	−1.1
1986–87	2.3	1.2	3.8
1987–88[b]	9.7	5.4	11.8
1984–88 (overall)[c]	9.7	8.5	14.2

[a]"Change" is percentage increase/decrease in *annual average* index for years covered by "period."
[b]Period covered is 1987 (annual average index) to 1988 (index for August or third quarter, whichever applies).
[c]Period covered is 1984 (annual average index) to 1988 (index for August or third quarter, whichever applies).

down in plant construction and competition from overseas fabricators. As Table 8.4 shows, the price picture has changed dramatically since then. The PPI also showed small decreases from 1984 to 1985, and in the 1985–86 period. Conversely, the M&S has grown steadily in the last half-decade. However, in the 1984–87 period, its increases were modest.

ESCALATING AGGREGATED CONTROL COSTS

In terms of magnitude and composition, aggregated control costs are completely different from equipment costs. Hence, they must be escalated differently. Because aggregated costs are generally less accurate estimates than equipment prices, one needn't be as particular in the selection of the index used to escalate them. The only requirement is that the index be broad-based and reputable. Two indices that meet those requirements are the Gross National Product Implicit Price Deflators ("GNP Deflators") and the Consumer Price Index ("CPI-U"). The GNP Deflators are compiled quarterly by the U.S. Department of Commerce's Bureau of Economic Analysis and published monthly in their *Survey of Current Business*. With a base of 1982 (= 100), the Deflators are computed for a composite ("Gross National Product") and for various categories, such as "personal consumption expenditures" and "gross private domestic investment." The composite Deflator would seem appropriate for the escalation of aggregated costs. Recent values for the composite are:[10]

Year	GNP Deflator (composite)
1986	113.9
1987	117.7
1988: 1st qtr.	119.4
2nd qtr.	121.0
3rd qtr.	122.4

The familiar Consumer Price Index is also handy for escalating aggregated costs. In fact, we used the CPI to escalate the costs in Chapter 1 to a common basis (1987 dollars). Recent average values for the CPI-U (the CPI for "All Urban Consumers," U.S. city average) are given below.[9] (*Note:* 1982–84 = 100.)

Year	CPI-U
1984	103.9
1985	107.6
1986	109.6
1987	113.6
1988: January–July	116.8
August	119.0

USING INDICES TO PROJECT COSTS

Indices are invaluable sources of *historical* price information. However, one may also be tempted to use indices to project costs into coming months and years. Resist that temptation! These projections, regardless of their mathematical sophistication, are usually worth very little. Not that heroic attempts haven't been made. One analyst regressed values for several indices (among them the CE composite) between 1971 and 1978.[11] From these regressions, he obtained two functions, a short-term (exponential) and a long-term (linear). Each function plotted the index value versus the month in which it was calculated.

Another analyst used these functions to predict values for the CE Index for three months, one each in 1979, 1980, and 1981. The observed differences between the actual and predicted indices were as follows:[12]

Date	Difference (%)
December 1979	–1.8
August 1980	–3.0
June 1981	–7.2

In all cases, the predicted index values were *less than* the actual. Note however, that the further we get from the 1971–78 regression period, the larger the difference between the indices becomes. This should tell us that such projections, if used at all, are of little value more than a few months in advance.

*   *   *

A few closing remarks seem appropriate. Nothing fancy. Just some thoughts to take with you when you close this book.

First, we sincerely hope that you *open* it again – and again, and again. For we think that it'll be well worth your while if you do. Although the book doesn't contain cost data on every air pollution control method in use today, it covers enough of them to make it a more-than-passable primer on APC cost estimating. Admittedly, the cost data do reflect a variety of dates. But via the procedures and data in this chapter, these costs can readily be adjusted to any desired date.

Specialists may claim that the "descriptions" and "sizing procedures" in Chapters 4 to 6 are superficial, for instance, while generalists might say that they're too detailed. However, we think we've provided just the right amount of information to enable the reader to make sound "budget-study" cost estimates.

This sort of "darned-if-you-do, darned-if-you-don't" criticism reminds us of what the late Al Capp once said about cartoonists, of whom he was one of the very best: "Artists look down on cartoonists because they're also writers, while writers belittle them because they're also artists." Rest assured that *this* author is neither! So please judge the figures and tables on their content, not their appearance. As for the text, kindly evaluate its problem-solving potential, not its literary merit.

Finally, if there's anything you like or don't like about this book (especially the latter) we want to know about it. "We" meaning this publisher and this author. The only way this (or any other) book can be made better is for its readers to supply honest feedback. Also let us know if you have any questions about using the book — or about cost estimating, in general. In our opinion, those who pay $40, $50, or more for a book deserve more than just a two-pound assemblage of paper and cloth. They deserve some help when they need it.

Good luck and happy estimating!

REFERENCES

1. "CE Plant Cost Index — Revised." *Chemical Engineering*, April 19, 1982, p. 153.
2. "Economic Indicators." *Chemical Engineering*. (Various issues.)
3. Stevens, R. W. "Equipment Cost Indexes for Process Industries," *Chemical Engineering*, November 1947, pp. 124–126.
4. Matley, J., and A. Hick. "Cost Indexes End 1987 on an Upswing," *Chemical Engineering*, April 11, 1988, pp. 71–73.
5. *The Producer Price Index: An Introduction to Its Derivation and Uses.* Washington: Bureau of Labor Statistics, 1988.
6. Price and technical data from W.W. Sly Co. (Cleveland, OH), February 1988.
7. "News-Record Indexes: History and Use." *Engineering News-Record*, September 1, 1949, pp. 398–432.
8. "Nelson Indexes Change Names." *Oil and Gas Journal*, October 5, 1987, pp. 84–85.
9. "Consumer Price Index — All Urban Consumers (CPI-U)." Washington: Bureau of Labor Statistics, 1988.
10. "National Income and Products Accounts Tables." *Survey of Current Business*, November 1988, pp. 3–18.
11. Mascio, N. E. "Predict Costs Reliably Via Regression Analysis," *Chemical Engineering*, February 12, 1979, pp. 115–121.
12. Vatavuk, W. M. "Control Costs." In: *Handbook of Air Pollution Technology.* New York: John Wiley and Sons, 1984, p. 372.

APPENDIX A

Depreciation and the 1986 Tax Reform Act

The 1986 Tax Reform Act drastically modified the federal income tax depreciation rules. Although some consider these new rules to be easier to understand, others think them more restrictive than those they replaced. This appendix highlights these new rules for the convenience of both categories of readers.

According to the Internal Revenue Service (IRS) publication *Depreciation*, depreciable property must: (1) "be used in business or. . .held for the production of income," (2) "have a determinable life longer than one year," and (3) "be something that wears out, decays, gets used up, becomes obsolete, or loses value from natural causes."[1] Depreciable property may be *tangible* (i.e., "anything that can be seen or touched") or *intangible* (e.g., a copyright or franchise). It may also be *real* (real estate) or *personal* property (machinery, equipment, etc.). Although buildings or other structures erected on land may be depreciable, land itself is *never* depreciable.

For every item of depreciable property, a depreciation tax deduction is calculated for every year that the property is in service. (In reality, a property may indeed be *in use* for longer than the allowable depreciation period, but IRS doesn't seem to mind.) The amount of this deduction is determined by the "modified accelerated cost recovery system" (MACRS) percentages. This system applies to ALL property placed in service after 1986. The MACRS divides property into six *classes* (3-, 5-, 7-, 10-, 15-, and 20-year property), each determined by its *class life*. Table A.1 lists class lives for selected equipment.

In figuring the depreciation deduction, two methods may be used: *straight line* (SL) or *declining balance*. In the *straight line* method, the amount of the deduction in year 1 is:

$$SL = TDI/m \qquad (A.1)$$

where $\quad$ TDI = total depreciable investment
$\quad\quad\quad\quad$ m = length of IRS-allowed "recovery period" remaining (years)

Some explanation is in order here. First, the TDI in Equation A.1 must be reduced to allow for any "Section 179" deduction applied against it. This deduction allows one to treat a capital expenditure as an expense and to deduct all or part of the full investment in the first tax year it is put in service.

Table A.1   IRS Table of Class Lives and Recovery Periods

Description of Assets	Class Life (years)	Recovery Periods (years)[a] General	Recovery Periods (years)[a] Alternative
General Depreciable Assets			
Office Furniture, Fixtures	10	7	10
Information Systems	6	5	5
Automobiles, Taxis	3	5	5
Light Trucks	4	5	5
Heavy Trucks	6	5	6
Specific Depreciable Assets			
Mining	10	7	10
Petroleum Refining	16	10	16
Manufacture:			
Grain and Grain Mill Products[b]	17	10	17
Tobacco and Tobacco Products[b]	15	7	15
Wood Products and Furniture	10	7	10
Pulp and Paper	13	7	13
Chemicals and Allied Products	9.5	5	9.5
Rubber Products	14	7	14
Finished Plastic Products	11	7	11
Cement	20	15	20
Primary Nonferrous Metals	14	7	14
Foundry Products	14	7	14
Primary Steel Mill Products	15	7	15
Fabricated Metal Products	12	7	12
Motor Vehicles	12	7	12
Electric Utility Steam Production Plant	28	20	28
Electric Utility Nuclear Production Plant	20	15	20
Waste Reduction and Resource Recovery Plants	10	7	10

Source: Depreciation, U.S. Internal Revenue Service, Publication 534 (revised December 1987), pp. 98-109.
[a]"General" refers to the "Modified Accelerated Cost Recovery System" (MACRS), "Alternative" to the *alternative* MACRS.
[b]Apparently, the IRS rules apply to Mother Nature also, for hers is the only firm that "manufactures" grain and tobacco.

However, the Section 179 deduction is limited to $10,000 for *all* investments in any given tax year.

Secondly, the IRS "straight line" method is a little different from the traditional SL method, in that the allowable deduction is *not* necessarily the same every year. When would it be different? Only if the property were placed in service during the tax year (e.g., in June), in which case the "half-year" convention would apply. Then, the depreciation deduction for the first tax year would only be *half* the full-year deduction.

Example: A piece of equipment costing $100,000 (its "basis") and having a 10-year "recovery period" is placed in service in mid-year. Via SL, the first-year deduction would be: $0.5 \times \$100,000/10 = \$5000$. In year 2, the deduction would be: $[\$100,000 - 5000]/[10 - 0.5] = \$10,000$; in year 3, $\$85,000/8.5 = \$10,000$, and so forth in every year except the 10th, in which the deduction would be $15,000. However, as

Table A.2   IRS-Allowed Declining Balance Rates for Property Classes

Class	Declining Balance Rate (%)	Year
3	66.67	3rd
5	40	4th
7	28.57	5th
10	20	7th
15	10	7th
20	7.5	9th

Source: Depreciation U.S. Internal Revenue Service, Publication 534 (revised December 1987), p. 6.

the convention used in this book specifies that all costs and revenues occur at the *end* of the years in question, all depreciation deductions will be figured on a full-year basis.

If the straight line method is chosen for any class of property, it must be used for *all* property of that class placed in service that year. Furthermore, SL must be used for every year of the recovery period.

Many firms elect to use the *declining balance* method in figuring depreciation. IRS allows the use of the double (200%) declining balance with property in the 3-, 5-, 7-, or 10-year classes, and the 150% declining balance method for 15- or 20-year class property. (*Note:* as might be deduced, the "class life" doesn't necessarily equal the "recovery period." See Table A.1.) For all classes, the taxpayer has to switch to the straight line method in the first year that the SL method would yield a larger depreciation deduction.

The *double declining balance* deduction (DDB) in year k is figured as follows:

$$DDB_k = 2/c \times TDI_k \qquad (A.2)$$

where
$$c = \text{class life (years)}$$
$$TDI_k = \text{adjusted basis at the start of year k}$$
$$= TDI - \text{sum of depreciation deductions in years 1 to } k - 1, \text{ inclusive}$$

Equation A.2 is also used to figure the 150% declining balance depreciation, except that "1.5" is substituted for "2" in the numerator. Table A.2 lists, for all property classes, the declining balance rates and years in which the taxpayer is to switch to the straight line method.

Lastly, one is also permitted to use the *alternate MACRS* method for most property. The catch is, the straight line method then must be used in figuring depreciation throughout the *entire* recovery period. No salvage is allowed either. Moreover, as Table A.1 shows, the alternate MACRS recovery periods are typically longer than the general depreciation system (MACRS) periods. For these reasons, the alternate MACRS would be most attractive to those firms who would benefit from the depreciation deductions more in the later years of the project than in the earlier.

Although the foregoing discussion has focused on the federal income tax rules, it applies also to the many states who have incorporated the IRS depreciation rules into their tax codes.

REFERENCE

1. *Depreciation*. U.S. Internal Revenue Service, Publication No. 534, December 1987.

Nomenclature

Following is a list of definitions of the terms used in this book, along with the units in which they are expressed. Where no units are listed, the term is mathematically dimensionless. Because we could devise no set of nomenclature that would "fit" all the terms in the book, we have instead developed a separate set for each chapter. Nonetheless, certain terms (e.g., "Q" for gas volumetric flowrate) are common to all chapters.

Chapter 2

A = annual capital recovery payment (\$/year)
a = correlation parameter
b = correlation parameter
C_{rp} = equivalent annual replacement parts cost (\$/year)
CRF = capital recovery factor
CRF_{rp} = capital recovery factor (replacement parts and labor)
DAC = direct annual cost (\$/year)
Dep = depreciation (\$/year)
$EUAC$ = equivalent uniform annual cost (\$/year)
FP = fan power (hp)
i = real annual interest rate
i_n = nominal annual interest rate
IAC = indirect annual cost (\$/year)
k = time from year "0" to year in which a cash flow occurs (years)
L = labor requirement
m = project life (years)
n = combined efficiency of fan and fan motor
NCF = net cash flow (\$/year)
$NDCF$ = net discounted cash flow (\$)
P_{rp} = cost of replacement parts and labor, including taxes and freight (\$)
ΔP = pressure drop (in. w.c.)
PW = present worth (\$)
Q = waste gas volumetric flowrate (acfm or scfm)
r = annual inflation rate

RC = recovery credits (\$/year)

ROI = return on investment

t = federal-state-local annual income tax rate

TAC = total annual cost (\$/year)

TAC^* = total annual cost less capital recovery charge (\$/year)

TCI = total capital investment (\$)

TDI = total depreciable investment (\$)

$TNDI$ = total nondepreciable investment (\$)

X = production level

Chapter 4

A = cyclone inlet area or building floor area (ft^2)

a = correlation parameter

b = correlation parameter

BHP = brake horsepower (hp)

D = fan wheel, duct, or damper diameter (in.)

d = duct diameter (ft)

d_c = critical particle size (μm)

d_h = circular hood diameter (ft)

E = heat transfer rate from hot source to plume (Btu/min)

H = stack height (ft)

H_s = height of structures in vicinity of stack (ft)

L = length of duct, hood, or screw conveyor *or* lesser dimension of structures near stack (ft)

N = fan speed (rpm)

n' = fan efficiency

P_i = atmospheric pressure at fan installation (psia)

P_s = standard atmospheric pressure (psia)

ΔP_d = duct static pressure loss (in./100 ft)

p = perimeter of hooded source (ft)

Q = gas flowrate (acfm)

r = gas density (lb/ft^3)

r_p = particle density (lb/ft^3)

SP = fan static pressure (in. w.c.)

T = temperature (°F)

t = absolute temperature (°R)

Δt = temperature difference between hot source and ambient air (°F)

TP = total fan pressure (in. w.c.)

u = gas viscosity (lb/ft-sec)

v = duct or hood capture ("face") velocity (ft/min)

v_i = cyclone inlet velocity (ft/min)

VP = fan velocity pressure (in. w.c.)

W = duct weight (lb)

w = rectangular hood width (ft)

w' = width of hooded source (ft)

z = open distance between source and hood (ft)

z' = altitude above sea level (ft)

Chapter 5

A = electrostatic precipitator collecting area or venturi scrubber throat area (ft^2)

A (gross) = gross filtering bag area (ft^2)

A (net) = net filtering bag area (ft^2)

a = correlation parameter

b = correlation parameter

C = constant in venturi throat equation

C_{br} = bag replacement cost ($/year)

CRC = capital recovery cost ($/year)

CRF_b = capital recovery factor for replacement bags

CRF_s = capital recovery factor for control system

D = mist eliminator diameter (ft)

D_c = absorption column diameter (ft)

d_{50} = mass median particle diameter (μm)

d_{rc} = particle cut diameter (μm)

E = overall particle collection efficiency

f = factor to account for extra bag area

g_b = scrubber bleed rate (gal/min)

H_a = absorption column added height (ft)

H_c = absorption column height (ft)

H_r = absorption column removal section height (ft)

h = pump liquid head (ft of water)

HTU = height of each transfer unit (ft)

L/G = scrubber liquid/gas ratio (gal/1000 ft^3)

l = pump liquid flowrate (gal/min)

M_g = gas mass flowrate through absorption column (lb/min/ft^2)

M_{in} = inlet particulate loading (lb/min)

M_w = water mass flowrate though absorption column (lb/min/ft^2)

n' = fan efficiency

n* = pump efficiency

NTU = number of transfer units

P_b = price of full set of bags, including taxes and freight ($)

P_{bl} = labor cost to replace bags ($)

P($) = equipment price ($)

PE = purchased equipment cost ($)

ΔP = pressure drop through venturi throat (in. w.c.)

p_{pl} = density of material of construction (lb/ft^3)

p_t = overall mass penetration
Q = volumetric flowrate (acfm)
R = gas-to-cloth ratio (ft/min)
r_g = gas density at saturation (lb/ft^3)
s = allowable solids content in recirculating scrubber water (lb/lb water)
s.d. = standard deviation of particle size distribution (μm)
TAC = total annual cost ($/year)
TCI = total capital investment ($)
V = recirculation tank volume (gal)
v_m = precipitator effective migration velocity (ft/min)
v_r = absorption column removal velocity (ft/min)
v_t = venturi scrubber throat velocity (ft/sec)
W_c = absorption column shell weight (lb)
x_b = mass fraction of solute in liquid at column bottom
x_t = mass fraction of solute in liquid at column top
y_b = mass fraction of solute in gas at column bottom
y_t = mass fraction of solute in gas at column top

Chapter 6

a = correlation parameter
b = correlation parameter
C_c = catalyst replacement cost ($/year)
C_{cb} = carbon replacement cost ($/year)
C_p = mean heat capacity (Btu/scf-°F)
C_{RTO} = installed cost of RTO ($)
CRC_s = capital recovery cost for catalytic incinerator or carbon adsorber system ($/year)
CRF_c = capital recovery factor for catalyst
CRF_{cb} = capital recovery factor for adsorber carbon
CRF_s = capital recovery factor for catalytic incinerator or carbon adsorber system
D_t = flare tip diameter (in.)
E = fraction of inlet adsorbate removed during adsorption cycle
F_{ef} = fuel gas requirement for elevated flare (million Btu/hr)
F_g = gross RTO fuel requirement (Btu/hr)
F_n = net RTO fuel requirement (Btu/hr)
F_{gf} = fuel gas requirement for ground flare (million Btu/hr)
f = fraction of flare flame heat radiated
G = cooling water usage (gal/year)
H = annual operating hours (hr)
h = heat content of waste gas (Btu/scf)
k = allowable heat flux from flare flame (Btu/hr-ft^2)
l_b = carbon bed thickness (ft)

LEL = percent of lower explosive limit

M_v = molecular weight of VOCs (lb/lb-mol)

m_e = equilibrium adsorptivity (lb adsorbate/lb carbon)

m_w = adsorber working capacity (lb adsorbate/lb carbon)

N_a = number of adsorbing carbon beds

N_d = number of desorbing carbon beds

P_c = initial price of catalyst, including taxes and freight ($)

P_{cb} = initial price of system carbon, including taxes and freight ($)

P_{cbl} = labor cost to replace carbon ($)

P_{cl} = labor cost to replace catalyst ($)

P($) = equipment price ($)

ΔP_b = carbon bed pressure drop (in. w.c.)

PE = purchased equipment cost ($)

PT, I, and AC = property taxes, insurance, and administrative charges ($/year)

p = partial pressure of adsorbate in waste gas stream (psia)

Q = waste gas volumetric flowrate (scfm)

Q' = inlet waste gas flowrate (actual gal/day)

Q'' = waste gas mass flowrate (lb/hr)

q = sensible or latent heat removed by refrigeration unit (Btu/hr)

R = condensate recovered by refrigeration unit (lb/year)

S = steam usage (lb/year)

T = temperature (°F)

t_a = adsorption time (hr)

t_d = desorption time (hr)

TAC = total annual cost ($/year)

TCI = total capital investment ($)

v_b = superficial bed velocity (ft/min)

v_e = flare gas exit velocity (ft/min)

W_c = carbon requirement, continuously operated adsorber (lb)

W_{ci} = carbon requirement, intermittently operated adsorber (lb)

w_{ad} = adsorbate inlet loading (lb/hr)

w_c = VOCs condensed in adsorber (lb/hr)

y = waste gas saturated VOC content (lb-mol/lb-mol air)

Z = distance from flare flame center to nearest point of allowable radiation intensity (ft)

& = heat exchanger efficiency

$\&_{RTO}$ = thermal efficiency of a regenerative thermal oxidizer (RTO)

Chapter 7

A = storage pile level land area (acres)

a = correlation parameter

  b = correlation parameter
CRC = capital recovery cost ($/year)
CRF = capital recovery factor
  i = annual interest rate
  k = constant, value of which depends on coal preparation level
M_{raw} = raw coal mass flowrate (tons/hr)
M_{ref} = coal refuse mass flowrate (dry tons/hr)
M_{we} = thermal dryer water evaporation rate (tons/hr)
  S = storage pile surface area (ft^2)
TAC = total annual cost ($/year)
TCI = total capital investment ($)
 WC = working capital ($)

Appendix A

   c = IRS class life (years)
DDB_k = double-declining balance depreciation deduction in year k ($)
  m = length of IRS-allowed "recovery period" remaining (years)
 SL = straight-line depreciation deduction in first year ($)
TDI = total depreciable investment ($)
TDI_k = total depreciable investment at the start of year k, "adjusted basis" ($)

Index